The Fate of the Peasantry

Premature 'Transition to Socialism' in the Democratic Republic of Vietnam

Andrew Vickerman

Monograph Series Number 28 Yale University Southeast Asia Studies
Yale Center for International and Area Studies

The Fate of the Peasantry

International Standard Book Number: 0-938692-26-7

Library of Congress Catalog Card Number: 86-40291

Distributor:
Yale University Southeast Asia Studies
Box 13A Yale Station
New Haven, Connecticut 06520

Printed in the U.S.A.

CONTENTS

List of Figures vii

List of Tables viii

List of Appendix Tables x

Abbreviations xi

Preface xiii

Introduction 1

Chapter I. A Background to the DRV

The Physical Background 24
The Impact of Colonialism 25
An Overview of the Development of Economy and Industry 1954-75 32
Political Background 41

Chapter II. From Colonialism to Partition: Rural Policy Developments, 1939-53

Introduction 45
The Japanese Interlude 47
The Anti-Colonial Struggle, 1945-53 49
Summary 70

Chapter III. The Agrarian Reforms and Their Aftermath, 1953-57

The Timing of the Agrarian Reforms 73
Early Phases of the Rent Reduction Campaigns 77
The Land Reform Campaign 81
Class Classification 83
Admission of 'Errors' and Rectification 98
Land Distribution and Landholdings 104
Summary of the Land Reform Period, 1945-57 113

Chapter IV. Launching Collectivisation

Introduction 117
The DRV's Own 'Scissors Crisis' 119
First Steps--Mutual Aid Teams 126
Collectivisation 130
Summary 156

Chapter V. Emerging Problems and Policy Responses, 1961-65

Introduction 159
Cooperatives, Brigades, and Families 160
The Early Years 166
Some Policy Responses 176
The Technical Development of Agriculture 190
Shifts in Focus, 1964-65 193
Summary 204

Chapter VI. Agriculture in a War Economy

Managing a War Economy 207
Decentralisation Under Siege Economy Conditions 208
The Old Approach Reinstated, 1968-70 218
Development in Adversity, 1970-73 233
Planning for Peace-Time Agriculture 237
The Thai Binh Conference 248
Summary 256

Conclusion

Summary 259
Postscript: A Vietnamese Reappraisal 264

Appendix Tables 273

Notes 293

Vietnam Bibliography 345

Supplementary Bibliography 367

FIGURES

1. Rice and Staple Food Production, 1954-75 38

2. Changes in Gross Value of Industrial and Agricultural Output, 1956-75 39

3. Index of Gross Value of Output--Agriculture and Industry & Handicrafts, 1955-75 40

4. Index of Gross Value of Output--Departments A and B, 1955-75 40

5. Index of Gross Value of Output--Industry and Handicrafts, 1955-75 40

6. Index of Gross Value of Output--Central and Local Industry, 1955-75 40

7. Index of Gross Value of Consumer Goods Output 40

8. Index of Gross Value of Output, Capital, and Workforce--Industry and Handicrafts, 1955-75 40

TABLES

1.1	Landownership in Tonkin in 1930	30
2.1	Land Appropriation in the DRV, 1945-57	61
2.2	Class Proportions and Landownership in the DRV, 1945-53	63
2.3	Class Proportions, Landownership, and Use per Family in the DRV, 1945-53	65
2.4	'Rich' Peasant Landownership and Use in the DRV, 1945 and 1953	68
2.5	Relative Per Capita Landownership in the DRV, 1945 and 1953	70
3.1	Waves of Rent Reduction and Land Reform	82
3.2	Renting Out Land and Hiring-in Labour, 1945 and 1949	86
3.3	'Landlord' Families, 1945-53	87
3.4	Exercise in Land Distribution	91
3.5	'Landlord' and 'Rich' Peasant Families, 1945-53	93
3.6	Land Owned and Used Before and After Land Reform	107
3.7	Rural Population During Land Reform and Land Held, 1956	109
3.8	Production During and After Land Reform, 1954-59	111
4.1	State Share of Wholesale and Retail Trade, 1954-60	120
4.2	Price Movements, 1955-59--Indexes	121
4.3	State Share of Paddy Crop (Taxes and Purchases), 1955-60	122

4.4	Paddy Production by Harvest, 1955-60	123
4.5	Peasant Families in Agricultural Organisations, 1955-60	128
4.6	Rural Structure, March 1959	136
4.7	Income Per Head, 1957-61	137
4.8	Peasant Family Cash Reserves, March 1959	139
5.1	Index of Livestock Numbers, 1956-64	168
5.2	Agricultural Production and Procurement, 1961-65	175
5.3	Investment in Agriculture and Irrigation, 1955-71	191
6.1	Agricultural Production, 1966-75	212
6.2	Peasant Income, 1965-75	224
6.3	Spring Rice, 1965-75	236

Missing data in tables is represented by the following symbols, which reflect standard United Nations usage:

....	Data not available
----	None, in negligible quantity, or not applicable

APPENDIX TABLES

I	Change in Landownership and Use, 1945-53	273
II	Classification and Landownership, 1945-56	276
III	Collectivisation of Agriculture, 1955-75	278
IV	Agricultural Production, 1954-75	279
V	Livestock Numbers, 1955-75	281
VI	State Share of Food Production, 1955-75	283
VII	Growth and Structure of Gross Value of Agricultural Output, 1955-75	285
VIII	Value of Labour Productivity	286
IX	Productivity in Cooperatives	287
X	Peasant Income Data, Selected Years	289
XI	Industrial and Handicraft Production Indexes, 1955-75	290

ABBREVIATIONS

CC	Central Committee
CPSU	Communist Party of the Soviet Union
CPV	Communist Party of Vietnam
DRV	Democratic Republic of Vietnam
H	Hanoi Radio
ICP	Indochinese Communist Party
MAT	Mutual Aid Team
NCNA	New China News Agency
NMS	New Management System
SRV	Socialist Republic of Vietnam
SWB	Summary of World Broadcasts
VNA	Vietnamese News Agency
VWP	Vietnam Workers' Party

PREFACE

Although it is called socialism, literally it is small-scale agricultural production.[1]

This comment was made in 1975 by Le Duan, First Secretary of the Vietnam Workers' Party (VWP), at a time when, according to official statistics, 93% of all peasant families and 89% of the cultivated area in the Democratic Republic of Vietnam (DRV) were in high-level cooperatives. In these cooperatives land and means of production were held in common. Income was distributed on the basis of labour input and all aspects of production were planned.

The policy package adopted in the DRV was based on Soviet and Chinese experiences. It placed priority on industrialisation in which the focus was on heavy industry. The role of agriculture was to provide the resources for this development. Given the initial small size of the industrial sector and the absence of unlimited foreign aid, cooperatives, introduced on a large scale from 1959, were a means to 'tax' agriculture.

This study presents an economic history of the DRV agrarian sector. It shows that the discrepancy between principle and practice noted by Le Duan was the result of the adoption of a policy package inappropriate to conditions in the DRV. These conditions also prevented the VWP leadership from imposing its will regardless of peasant opposition. Peasant support for the resistance war against the French was gained by the promise of land. The 1953-56 agrarian reforms honoured this promise but also generated conflicts, which were inevitable in any attempt to create an egalitarian land ownership structure in a situation of extreme land scarcity. The radical Land Reform provoked economic and political opposition from the numerically significant group of peasants with above average land holdings.

---

1. H 31/1/75, SWB WS No. 814, p.A24.

Collectivisation encountered opposition of a more passive form than that expressed during the latter phases of Land Reform, as the peasantry continually reasserted the practice of family-based production. This tendency was stimulated both by the failure of cooperatives to demonstrate any material superiority over family production and the State's need to obtain produce from the family sector. The State's dependence on family production was inevitable given the limited surplus, unless cooperative output <u>and</u> productivity grew rapidly. The tendency for it to stagnate increased dependence on the family sector. The continued existence of the family sector was thus a compromise at the political and economic level.

During fifteen years or more between the initial development of cooperatives and Le Duan's remark, the VWP leadership continued to assert the principle of collective production and to attribute its failure to ineffective policy implementation, while the peasantry continued to focus attention and, wherever possible, resources on family production.

At a macro level the economy was in a vicious circle. The State lacked the ability to impose draconian taxation on agriculture and thus also relied on trade. However, the focus on heavy industry and the limited availability of consumer goods constrained the expansion of agriculture and its marketed surplus. The slow growth of agriculture, especially in the face of a rapid population growth, in turn constrained the development of industry.

This study focusses on some failures and problems of development in the DRV to provide a basis for analysis of subsequent measures in unified Vietnam, and for 'socialist' policy choices in other poor, labour-surplus economies. However, very <u>real</u> successes were achieved by the DRV, especially in health and education, as well as reunifying Vietnam and defeating the United States in a war that had a considerable impact on the economy. These successes are not ignored out of any desire to denigrate, but rather in the belief that a serious understanding of what went wrong is more useful than mere appreciation of what went right.

The study relies almost entirely on material available in the English language, much of it in translation. (The exceptions to this are certain statistical sources.) In the case of the BBC material there is also a dependence on what the DRV chose to braodcast and what the BBC chose to monitor, translate, and reproduce. Wherever possible, however, information has been cross-checked.

Although reference is made to the ideology and policy package 'inherited' from China and the Soviet Union, the experiences of these countries are not considered in great detail. The absence of extensive comparative material from other East Asian labour-surplus economies following different paths of agrarian development is a further omission. In part, justification for this omission is that the economic history of the DRV agrarian sector had to be written first and the space this required precluded detailed consideration of the experiences of other 'socialist' and 'non-socialist' economies. This leads to a further weakness in that the economic analysis is largely implicit in the economic history and may also appear somewhat fragmented and repetitive.

The study examines the agrarian sector throughout the period of the DRV's independent existence. This has both advantages and disadvantages. One disadvantage of such an approach is that it can take only limited account of industrial development except insofar as it impinges on agriculture. It also means that certain aspects of agricultural development are dealt with only briefly and others not at all (e.g. State farms and highland development); and, it excludes consideration of post-1975 developments. Post-1979 developments are considered briefly in the postscript since they relate closely to the critical view of collective agriculture presented here.

A comprehensive <u>economic</u> analysis of the DRV and Vietnam as a whole is long overdue. It is hoped that this agrarian economic history contributes to this task by filling an existing gap in the literature.

Financial support for the research was provided primarily by the Social Science Research Council (UK), but Queens' College Cambridge and the School of Development Studies at the University of East Anglia, where I was employed during the latter stages of the research also provided support. I would also like to thank my family, friends, and colleagues for support over the years. I am particularly grateful to Ken Post who commented on an early draft and provided me with drafts of his forthcoming major work on Vietnam.

I am most indebted to Suzy Paine and Lesley Brennan. Suzy stimulated my initial interest in the Vietnamese economy and helped and encouraged me throughout the writing of this book. Her 'fierce and unrelenting' criticism of earlier drafts made the final version substantially better than it would otherwise have been. Her death is a tragic loss to the all too thin cadre of serious critical analysts of socialist economic development in Asia.

Lesley's contribution is greater than I can easily acknowledge. Not only did she type and edit numerous drafts, but she provided the support and encouragement to sustain me through the various trials and tribulations of completing the Doctoral Thesis on which this book is based.

The book is dedicated to Justin and Jody who will one day understand why I spent so much time with books that have no pictures.

INTRODUCTION

An important focus of this book is on the rationale for collectivisation in the DRV, although its major purpose is to provide an economic history of the agrarian sector for the period 1945 to 1975. The study considers the three main arguments adduced by the DRV and VWP leadership for collectivisation and the evidence for their validity, both in the late 1950s when collectivisation began, and subsequently. This examination leads to the conclusion that collectivisation was inappropriate in the conditions of the DRV and failed to fulfill even the objectives laid down for it by the State.

The reasons for its inappropriateness and relative failure are seen in the policy framework itself as well as in the conditions existing in the DRV. In other words, it is tentatively asserted that an expansion of agricultural production, both absolutely and per capita, would have required a different institutional framework for agriculture as well as different overall economic conditions. On the more limited criteria of satisfying the objectives laid down by the State it could be argued that the basic conditions were unsuitable.

The rationale for collectivisation in the DRV was not original; similar arguments had been used by the Soviet and Chinese leadership. This introduction considers aspects of the theoretical background for these arguments as well as evidence for their validity in the cases of the USSR and China.

Implicit in the critique of collectivisation are elements of an alternative institutional framework for agriculture, namely, one based on individual peasant household cultivation. Indeed, certain characteristics of peasant household behaviour and decision-making are actually seen as a reason for the relative failure of cooperatives. This study does not attempt to suggest the characteristics of any alternative or to discuss 'models' from other countries, although it does suggest that more appropriate models than the Soviet Kolkhoz do exist for Vietnam.[1]

Therefore, this introduction also briefly discusses certain characteristics of the peasant economy in relation to the wealth of historical and theoretical material on the subject, although, no attempt is made to construct a complete model of the Vietnamese peasant economy.[2]

The overall framework in which the policy of and arguments for collectivisation must be seen, followed communist orthodoxy. China's strategy has been described as "a Stalinist strategy of economic development with local adaptations."[3] Accordingly, the DRV's strategy of economic development could be described as Stalinist with some Maoist and local adaptations.

> The outstanding feature of the Stalinist model is the primacy of its concern with growth or, more specifically, with industrial growth.[4]

> The characteristic nucleus of the model may be identified...by the critical significance that is attributed to the extraction of a sizeable agricultural surplus of some sort in support of rapid industrialisation.[5]

A further characteristic is that collectivisation would be the means to obtain this surplus.

> If there were no other powerful reasons for the desirability of collectivisation of agriculture, the vital need for the mobilisation of the economic surplus generated in agriculture would in itself render collectivisation finally indispensable. By transferring the disposal of agricultural output from individual peasants to government-supervised collective farm managements, collectivisation destroys the basis for the peasants' resistance to the 'siphoning-off' of the economic surplus.[6]

One argument for collectivisation in the DRV was precisely that it would allow the State greater control over output and surplus and hence over the allocation between consumption and investment.

The other arguments or "powerful reasons for collectivisation" were that it would reduce peasant differentiation and prevent the emergence of capitalist agriculture as well as permitting a more rapid growth of production, apparently through economies of scale and a more sophisticated division of labour.

The ultimate genesis of the arguments will not be explored in great detail here. The focus on industrialisation derived from recognition of the backwardness of the USSR in its early years and the perceived necessity to 'catch' up with the capitalist industrialised economies, as well as from the notion that a socialist economy is an industrialised economy. The extent of the priority to industrialisation was the subject of debate among Soviet economists in the 1920s.[7] They were also concerned with the means of extracting the surplus from agriculture and the relative merits of taxation and use of the terms of trade in the absence, at that time, of "Stalin's re- markable far-sightedness"[8] and the option of forced collectivisation. Thus, the argument that agriculture should 'finance' industrialisation[9] has always been accepted among communist theore- ticians, while the extent to which agriculture should be subordinate to industry was still debated in the 1920s. Forced collectivisation was Stalin's unique contribution.

Preobrazhensky argued that the essential need was to modernise industry on the grounds that its growth potential was greater than that of agriculture. In an economy still dominated by agriculture this inevitably meant that agriculture would have to finance State industrialisation.

> To achieve this, prices charged by the State for the products of its industry should be such as would compel peasant purchasers to contribute to investment in the socialised sector.[10]

He termed this process of exploitation of the peasantry by the State the "Law of Primitive Socialist Accumulation."[11] He saw it essentially as a short-term sacrifice which would ultimately benefit the peasantry as the income of the whole of society rose.

Bukharin advocated a different approach with reduced emphasis on industrialisation, and with surplus extraction through taxation rather than through the terms of trade. He argued for the encouragement of increased accumulation in agriculture which constituted the demand for the products of industry. The rising incomes of the peasantry would be taxed to provide resources for industrial accumulation.[12] He argued that industry would decline if it did not develop on the base of a rapidly expanding agriculture.[13] This point--although not true for the USSR (until possibly the late 1950s) was true for China and the DRV.[14]

Both Preobrazhensky and Bukharin accepted that agriculture would have to finance industrial accumulation. Essentially, Preobrazhensky argued for maximum extraction in the short-term to generate rapid growth of industry which would then help to develop agriculture. A contradiction existed in Preobrazhensky's argument in that he accepted that the increase in the peasant's propensity to consume was the most important single stimulant to non-autonomous investment. The time periods involved are probably crucial to his formulation. Bukharin contended, rightly, that Preobrazhensky's 'superindustrialisation' programme would have a long gestation period. The cost of limiting consumption and its rate of growth in order to maximise accumulation might constrain the ability to generate resources for accumulation in the future. The static choice between consumption and accumulation ignores certain important dynamic considerations.

Bukharin argued for extraction from rising peasant incomes which involved encouraging agricultural accumulation. Preobrazhensky called this "riding into socialism on a peasant nag."[15] This expression illustrates Bukharin's view as well as the disparaging view of the peasantry shared by much of the urban Bolshevik leadership. As is shown later, the DRV leadership held a similar view of the Vietnamese peasantry.

Bukharin and Preobrazhensky had different opinions about the most appropriate means of extraction--namely through manipulation of the terms of trade or through taxation.[16] They differed also concerning the time periods involved. Stalin's contribution was to give greater priority to industry for a longer period. For any foreseeable planning period "steel was a final good to Stalin, and bread an intermediate one."[17] He also employed a process of direct extraction through government and party-led collectives.

This summary of the debate in the USSR is necessarily brief. The important point is that the same argument that agriculture could finance industrialisation with cooperatives as a means of gaining access to marketings and surplus, was repeated both in China and in the DRV. Agriculture was collectivised in order to collect tribute, which Preobrazhensky, Trotsky and--to an extent, Bukharin--had argued was required. Yet, none of these advocated forced collectivisation, or "primitive socialist accumulation by the methods of Tamerlane."[18]

However recent work[19] has revealed that in the case of the USSR,

> during the First Five Year Plan (1928-32) there was no increase in the net transfer of commodities to industry....Hence it would appear that for the USSR in 1928-32 the view that collectivisation enables the net transfer of commodities from agriculture to be increased is wrong. An important argument for collectivisation appears to be fallacious.[20]

While agriculture made a contribution by providing an increased supply of foodstuffs (i.e. the marketed surplus rose), and much of the labour force for industry,[21] as well as exports and imports,

> there is no basis whatsoever for the view that the increase in investment during the First Five Year Plan was financed by an increase in the agricultural surplus.[22]

Accumulation was actually at the expense of

urban workers with a 40% decline in real wages between 1929 and 1923.[23] Trotsky had actually foreseen the need to extract 'tribute' from the working class to finance industrialisation.[24] It appears that while the State did turn the terms of trade in its favour for 'official' transactions, the continuation of substantial free market transactions with rapidly rising prices meant that the actual terms of trade for many peasants did not decline. In such circumstances, also apparent in the DRV, peasants will endeavour to sell on the free market and buy on the official market.

The same argument of cooperatives as a means of surplus extraction was also employed as a rationale for collectivisation in China[25] and has similarly, although more tentatively, been shown to be fallacious.[26] This study will show that the argument was repeated in the DRV and appears to have been similarly fallacious.[27]

Two questions become obvious: first, did it matter that little "primitive socialist accumulation" occurred at the expense of agriculture[28] and second, why didn't it occur when fifty years of communist theory thought it both would and did?[29]

The answer to the first question seems to be that it did not matter in the USSR but did in China and the DRV.

> The Soviet Union achieved growth and prosperity not because the peasantry was exploited and agriculture neglected, but despite it.[30]

Answering the second question involves examining the other arguments for collectivisation and considering their validity. The basis of the Preobrazhensky, Stalin, and subsequent arguments was the belief in the need for an unconstrained rate of industrial expansion. It has been argued that in the case of the USSR, the industrialisation programme of the 1930s was unconstrained, whereas in the case of China it was constrained.[31] This resulted both from the initial differences in the base from which the economies developed in 1917 and 1949 and the levels of agricultural output per capita. China's per capita food availability was less than half that of the USSR at comparable dates.

This starting point allowed the USSR to obtain the desired level of marketings despite the calamitous falls in output of the early collectivisation years--the peasantry were the residual claimants on output.[32] Suggested tests of the existence of a constraint are the extent of correlation of industrial and agricultural growth rates, and the relative fluctuations of State procurement and farm consumption.[33] The absence of any significant correlation between sectoral growth rates, the small coefficient of variation of the industrial growth rate, and the greater fluctuations in farm consumption in the USSR imply that industry was unconstrained. In China, the growth rates are correlated, with a lag; there is extreme variability in industrial growth rates, and the State finds itself as a residual claimant on agricultural output.[34]

In the case of the DRV, Figure 2 clearly shows the correlation of growth rates, again with a lag, and fluctuations of industrial growth, even in those years unaffected by US bombing. The limited data on procurement make it difficult to draw any firm conclusion but it appears to resemble China more closely than the USSR. This would be expected given the low and falling level of per capita production in the DRV.[35] One minor point to note here is the difference between State grain extraction (as a proportion of the harvest) in the DRV as opposed to the other two countries. State procurement averaged 29% in the USSR for 1930-2,[36] 28% in China for 1953-4 to 1956-7,[37] and 20% in the DRV for 1960-5.[38]

Thus, the motivation for collectivisation was to control agricultural output in order to finance industrialisation and prevent agriculture from constraining industry. The possibility of industry constraining agriculture by failing to provide modern inputs and consumer goods is never considered in the classic Soviet model. It was, however, recognised by Bukharin and is discussed in this study. Chinese and Vietnamese policy in recent years show increasing recognition of this constraint.

It appears that agriculture did not finance industrailisation in any of the three cases and yet in the case of the USSR, neither did it act as a constraint. However, it is possible that collectivisation, by its contribution to the marketed surplus, may have actually prevented agriculture from acting as a constraint in the USSR. In China the direction of intersectoral resource transfer (from industry to agriculture) was

> brought about by agriculture's relatively slow growth during the 1950s to an extent that made it an increasingly effective constraint on growth in the rest of the economy.[39]

A similar case can be made for the DRV in the late 1950s and 1960s.

The question--why did agriculture grow slowly?--must therefore be answered and this involves consideration of a second major argument for collectivisation. It was argued in all three cases that cooperatives were, or would be, superior to private household agriculture in terms of output and productivity.[40] While this was asserted in all cases, in classical Marxism it depended on mechanisation first so that agriculture would closely resemble large-scale industry.

> Societal development in agriculture is taking the same road as in industry.[41]

In the Stalinist version, mechanisation would go hand-in-hand with collectivisation,[42] while in the Chinese (Maoist) version, accepted by the DRV, collectivisation should precede mechanisation.

> The relationship between mechanisation and collectivisation was now reversed; collectivisation became a prior condition of mechanisation.[43]

The orthodox and Stalinist view is based on the notion of economies of scale being related to mechanisation--the scale of production inevitably increases as industrialisation and mechanisation increase and the division of labour advances. Only large-scale mechanised collective farming could maximise agricultural production.[44]

Whether or not this view is valid--and the experience of the USSR would suggest it is not--it was further amended by Mao Tse Tung. This version, which was accepted by the DRV leadership, argued that as well as facilitating mechanisation, collectivisation culd partially substitute for it in the short-run by maximising the output and productivity potential of traditional inputs and technology. This greater attention paid to the means of raising agricultural production related to the constraint it posed for industrialisation. Something had to be done to remove or relax the constraint. Collectivisation was seen as having two potential advantages here. It could tighten the control over agriculture and it could allegedly raise output at minimal cost. This latter point was crucial--collectivisation would produce returns via exploitation of traditional technology at low resource cost.

> The new analysis made agricultural growth more a function of socialist organisation than the amount of industrial assistance available; institutional reform would permit technical advances which were largely independent of industry.[46]

This argument hinged on two assumptions: first, that an unexploited potential remained in traditional agrarian technology; and second, that only collectives could realise this potential.

> Altering the way in which labour was organised was not unimportant. Any positive gain from such a change, however, depended to a substantial degree on whether there was much room for increases in productivity left in the 'traditional' technology.[47]

Ishikawa argued, on the basis of a substantial amount of evidence on Asian agriculture, that different inputs are important at different stages of Asian agricultural development. Beyond the mere availability of land, sun, and rain, the important initial inputs are traditional investments in land. These 'basic investments' take the form of the provision of irrigation and water control facilities.

> In the later stages, a certain combination of improved farming techniques centering around improved varieties and increased application of fertilisers emerges as the leading input, still alternating with basic investment of an increasingly high quality.[48]

The crude cut-off points for these stages can be summarised as follows: basic conditions can produce paddy yields up to around 1.3 tons per hectare; basic land investment is the leading input between 1.3 and 2.3 tons per hectare; improved techniques are the key between 2.3 and around 3.8 tons per hectare. Beyond this, mechanisation plays the key role along with continued basic investment of a higher quality, and technique improvements.[49]

Ishikawa, and other work,[50] would suggest that China had reached the end of the stage in which basic investments were crucial by the late 1950s, if not earlier, and that the return to further use of traditional technology was rapidly diminishing.

> The most important source of agricultural growth during 1952-7 had been the increasing use of current inputs, mainly chemical fertilisers.[51]

If this argument is valid then the merit of cooperatives would lie in their greater suitability with new inputs. However, if it could be demonstrated that cooperatives could further exploit traditional technology or effectively ignore the rapidly diminishing marginal returns, then they would be important even with traditional technology. In other words, either Ishikawa's analysis may not be relevant to large-scale agriculture or its implications can be ignored. The rapidly diminishing returns are irrelevant if the objective is only output increases and if these are actually achieved. This foreshadows discussion of the conflicting objectives of maximising total, average, or marginal return.[52]

The Maoist argument focusses on the greater potential of cooperatives with both traditional and modern technology.

> Collective agriculture was presented as the only way for the peasant community as a whole to raise output and living standards, via labour-intensive work in the short-term, and the collective purchase of modern inputs in the long-run.[53]

The basis for this assertion was that economies of scale existed even with traditional technology. The possible sources of these economies were: more rational field layout, greater specialisation of both labour and land-use, irrigation improvements, and more advanced management of resources. The latter is included since, although a 'modern' input, it was arguably a low-cost one. The contributions of these potentially positive sources of output growth are difficult to measure, although as stated above, the potential return to large-scale labour mobilisation was limited.

It also seems clear from available evidence on the USSR, China, and the DRV that there were substantial negative effects of collectivisation. These relate mainly to the existence of diseconomies and the attitudes of the peasants themselves. There is an overlap here. If substantial economies of scale had existed, then the incentive for peasants to participate enthusiastically would have been greater. In the same manner, operation in the area of rapidly diminishing returns (for example to labour-intensive land investment) would increase any disincentive effect.

The absence of any evidence for the superiority of collective production in the DRV reinforced the peasantry's lack of enthusiasm for collective production. This in turn made it more difficult to establish any superiority.[54]

It should also be added that with the exception of the few instances referred to above, there is no clear evidence for economies of scale with either traditional or modern agricultural inputs, with the possible additional exception of extensive mechanisation.[55] The latter is of limited relevance given both the current state of Chinese and Vietnamese agriculture and the large amount of labour available. Paradoxically, the problems of labour management and organisation in cooperatives may actually have encouraged a cadre preference for more capital-intensive production techniques.[56]

Despite differences of opinion over whether collectivisation comes before, during, or after mechanisation, Marxist theoreticians have consistently believed in the merits of large-scale organisation. The following quotation from the founding fathers can be used to criticise at least the Maoist argument and to illustrate the underlying 'theoretical' belief in scale.

> The setting up of a collective economy is predicated upon the development of machinery, upon the utilisation of natural resources, and of many other productive forces....In the absence of these conditions, the collective economy would not in itself represent a new productive force, would be lacking in all material basis, and would rest upon purely theoretical foundations.[57]

This 'theoretical' belief has been proven incorrect partly due to the fundamental differences between industrial and agricultural work work which make it inherently much more difficult to supervise and "supervision is a fundamental barrier to economies of scale in agriculture."[58] These differences are: that there is less specialisation in agriculture; that the quality of work is not apparent until the end of the production process; the spatial nature of farm work; the lack of uniformity of resources; the unpredictability of resources and the natural environment; and the seasonal nature of farm work.[59]

These technical difficulties of work ogranisation, management, and supervision have been compounded in the cases of China and the DRV by the shortage of adequately trained managers and technicians. In some cases, even when these capabilities were present, cadres were tempted to exploit their positions for private gain.[60]

These same characteristics of agriculture which make large-scale labour organisation problematic also impinge on the attitudes of the peasants or agricultural workers themselves. It is implicit in much of the critique of collectivisation presented in this study that output would have been greater with a system of household production, and that the peasantry had a marked preference for such a system, although inadequate provision of modern inputs and consumer goods was a constraint on any system. This point has only recently been clearly recognised.

> To improve agricultural production, we are trying to produce more industrial goods to sell to the peasants in order to encourage them.[61]

It has been asserted in the case of the USSR that:

> All available evidence indicates that the performance of collective and State farm workers on their household plots has been superior to their performance in the socialised sector. This higher level of incentive on private plots is reflected in the tendencies of collective farmers to minimise their effort on collective work, and to spend excessive time and effort on their private plots. In part, this discrepancy is the result of the fact that the worker's income per unit of labour input is greater, and the relationship between effort and reward is clearer, on the household plots than in collective production. However, the greater efficiency of the household sector can also be explained partially by the fact that there is no problem of supervision in private plot production. In the private sector, it

appears that unsupervised initiative --resulting from the clear relationship between individual effort and individual income--is sufficient to generate high levels of worker performance. Attempts to deal with the problems of managing and supervising labour in the collectivised sector have been unable to produce the same kind of initiative, and...the peasants have been remarkably effective in evading the restrictions placed upon them by managers.[62]

Similar points have been made with respect to China[63] and are made for the DRV throughout this book, although the available empirical evidence is insufficient to fully substantiate them. This volume also attempts to consider the theoretical arguments accounting for these observations.

The conflict between peasant and State objectives is an abiding theme in this study. One final argument used by the leadership to account for many of the tendencies noted in the quotation above, and hence the relative failure of cooperatives to exhibit material superiority, was the inherent characteristics and ideology of the peasant. Inculcation of a new 'socialist consciousness' would overcome these difficulties. This would help make maximisation of output the objective and ignore the diminishing returns. However, as has been argued by Engels among others, the peasantry would only be attracted by socialist forms when it was demonstrably in its material interest to do so.[64]

A discussion of agriculture and industry, of peasant and State interests involves broad generalisations. With respect to peasant attitudes and interests, this study emphasises the need to distinguish between different groups and to consider the range of variables influencing peasant household decision-making. Some comments should therefore be made here regarding theories of the 'peasant economy'. Peasant economy has a bearing on the issues of incentives, participation, collective-versus-household production, and differentiation within the peasantry. It also relates to the third major argument for collectivisation in the USSR, China, and the DRV, in addition to the control and extraction, and economies of scale arguments already discussed. This third argument was the need to reduce differentials among the peasantry and by so doing, to prevent the emergence of capitalism in agriculture and the rural economy.

In his work on the peasant economy in Russia,[65] Chayanov raised a number of points which are relevant to part of the previous discussion as well as to some of the analysis in this study. He rejected the orthodox Marxist argument that "small-scale peasant agriculture must inevitably give way to large-scale capitalist agriculture."[66]

> Chayanov saw the modernisation of traditional small farming as lying along neither a capitalist nor a socialist road, but as a peasant path of raising the technical level of agricultural production through agricultural extension work and cooperative organisation, at the same time conserving the peasant institutional framework of the family small-holding.[67]

However, he later argued for large-scale agriculture in conjunction with extensive mechanisation.[68]

Chayanov also had doubts about the viability of collective production because of the incentive problem which

> had been solved more flexibly by the cooperatives based on the small family farm, with its individuality intact, than by the artels.[69]

Underlying these views was his analysis of the factors influencing peasant household decision-making and the differences between this and capitalist reckoning. He particularly emphasised that the peasant family farm uses a subjective evaluation of the balance between consumption and labour, although this was not an original contribution.

> The equilibrium of the family labour farm is depicted as the outcome of a subjective balancing of a marginal increment in family consumption against a marginal change in the 'drudgery' of family labour application.[70]

Finally, Chayanov argued that peasant household socio-economic mobility was determined by a range of mainly demographic factors and that no clear process of differentiation leading to capitalist agriculture could be observed.

> Economic stratification due to these causes Chayanov called 'demographic differentiation' to distinguish it from differentiation attributable to a persistent and cumulative process of petty capitalist accumulation.[71]

The demographic cycle ensured that no peasant family could obtain a permanent position of superiority over others, though it might do so temporarily. The importance of considering socio-economic mobility within the peasantry is emphasised in this study, although it does not imply acceptance of a simple life-cycle hypothesis.

There are several limitations in Chayanov's analysis for the purposes of this work. The first is his recognition that his work had limited relevance for densely populated regions and areas where there is no market in land.[72] Secondly, Chayanov's focus was inevitably on peasant economy within a capitalist economy. However, his work could

> lend support to the worst kind of heresy: that the peasantry might have both social justification and the economic capability to coexist with socialism as it had done with capitalism.[73]

His analysis of how peasant agriculture could

compete with capitalist agriculture could certainly provide a basis for analysing the relative performance of collective and family sectors within socialist economies, although this is not attempted in any rigorous fashion here.

Thirdly, "Chayanov's theory of peasant farming remained essentially a static one."[74] Finally, he tended to generalise across the whole peasantry whereas it is important to recognise the different interests of different groups or strata. This is true even if his argument about the causes of apparent differentiation is accepted. On the basis of a simple life-cycle hypothesis, the response of a household to changes in the terms of trade or its decision in respect of labour allocation will still differ during the cycle.

Another problem with Chayanov's conceptualisation of the family labour-consumer balance

> is the inability to investigate the intra-familial relations, in particular the sexual division of labour in society.[75]

This is not explored in this work although the postscript notes the need to analyse it in the light of more recent developments.

A similarity does exist between certain aspects of Chayanov's 'peasant economy' analysis and some of the analysis in this study, without implying that he provides a model suitable for examing all aspects of peasant behaviour within the framework of a formally socialised economy. There is, of course, considerable disagreement over the basic validity of Chayanov's model and its usefulness in analysing Russian peasant economy during the first twenty years of this century.[76]

If two extremes are posed by a Chayanovian vision of a semi-idyllic world of peasant farmers and a Soviet Marxist vision of a rigidly differentiated peasantry ripe for collectivisation, then rejection of the second view at a given historical juncture--the DRV in 1959--does not imply acceptance of the first.[77]

Implicit in the consideration of the Vietnamese peasant household is the basic Chayanovian hypothesis

> based on the concept of the peasant farm as a family labour farm in which the family as a result of its year's labour receives a single labour income and weighs its efforts against the material results obtained.[78]

However, the formulation of a complete model is complicated by the existence of two sectors to which labour is allocated; the existence of differing production functions in the two sectors; differing terms of trade depending on whether produce is bought and sold on the free or official markets;[79] and the existence of a range of different maximisation objectives. With respect to the last of these, for example, the State may wish to maximise total output, the existing cooperative members may wish to maximise the average product of labour and hence the workpoint, while individual workers may wish to equate the marginal productivity of labour with the marginal disutility of work.[80]

The household makes decisions on the allocation of its labour between different crops on its own plot, on the brigade land, and in handicraft production. The cooperative makes decisions on the allocation of its own labour among crops and on the purchase of additional labour. These decisions are influenced by conditions external to the cooperative--the fixed quota and price, excess delivery price, free market prices, and the price and availability of industrial goods--and external to the family--the labour obligation to the brigade, cooperative income and its distribution, and market prices. In addition the family's labour/consumption balance must be considered. Conditions clearly occur through which the existence of the family sector constrains the collective sector and vice versa.[81]

For example, if the return is higher on family plots because of both higher productivity and free market prices, then more labour will be applied to family production (i.e. a higher level of drudgery would be accepted because the return in terms of the marginal satisfaction of the goods obtained in exchange is greater).[82] However, the amount of labour which could be applied to the family plot is limited by its size, falling marginal productivity, and obligations to the cooperative. Higher production would be elicited by higher returns, although production was constrained by the limited availability of consumer goods as well as the existence of cooperatives.[83]

This book does not attempt to formulate a complete model of a peasant economy within 'socialism' but indicates certain theoretical reasons for observed phenomena, such as the tendency to focus effort on the family plot and derive a disproportionate income from it. Aspects relevant to the analysis of peasant behaviour within the collectivised framework are therefore raised in this presentation of a largely descriptive economic history.

In light of these introductory comments it is clear that there are considerable similarities between the experiences of the USSR, China, and the DRV and between the issues involved. The analytical issues raised in this introduction --the validity of the collectivisation rationale, the agriculture/industry relationship, the conflict between State and peasant objectives, and the nature of peasant decision-making--are discussed throughout the book. This consideration of the experience of the DRV should produce a searching reassessment of these issues and of the most appropriate institutional framework for agriculture as well as the overall economic development model for other poor, labour-surplus, agrarian economies.

The evidence from the DRV would certainly call for a reconsideration of the following view.

> ...poor, overcrowded Asian countries are the most likely loci for successful transformations to collective agriculture, due to the fact that the advantages of collective agriculture in raising output and living standards here have a potentially

strong appeal to a majority of the peasant population.[84]

Chapter II considers the development of VWP policy during the resistance war against France. It shows the economic necessity for reliance on a united front policy and the constraints this placed on agrarian reform measures. The peasantry supported the VWP because it 'promised' land, and the rising aspirations of the peasants-- fuelled by increasing impoverishment as well as education and literacy campaigns--necessitated fulfillment of the promise. The land scarcity meant this could only be achieved by breaking the 'united front.' By 1953, the balance of forces internally and internationally shifted more in favour of the implementation of radical reforms and the ending of the 'united front,' which would also consolidate the VWP monopoly of power.

However, the cross-section land distribution had also changed by 1953 because of the temporary allocations of previous years and the dispersal of land by relatively large landowners. The resulting pattern included a large group of peasants with per capita landholdings a little in excess of the mean. Given the absolute land scarcity and backward agrarian techniques, even the mean was barely adequate to provide basic consumption. Any attempt to implement an egalitarian reform would thus inevitably create substantial rural conflict.

Chapter III looks at the agrarian reforms and shows that the difficulty, outlined above, of reconciling the demand for land with its limited supply, was worsened by the kind of reform pursued. This reform was based on an ambiguous class classification derived largely from China. It was based on the mode of exploitation of land and took only limited account of family labour/ consumption balances and changes in these over time. Indeed, they changed considerably from year to year given the incidence of births and deaths and low life expectancy. Implementation of the agrarian reforms (classification and land appropriation and distribution) relied on peasant activists from outside the villages and, to a large degree, on VWP members who were not associated with those classed as 'landlords' and 'rich' peasants. This generated additional conflicts within the peasantry since all those involved in resistance organisations could be viewed with suspicion because these organisations had included the rural rich.

The attempt at a fairly egalitarian distribution hit those whose holdings had risen above the mean in earlier years as well as those 'wrongly' classified--either mistakenly, or to fulfill 'percentages,' or to gain land. The conflicts over land which were generated made corrective action inevitable. Its timing in 1956 was determined by the cumulative impact of the reforms; by their actual completion; by the consequences for food production and the State share of continued conflict, since limited food aid was available; and by international events which provided ideological justification for action to deal with the conflicts. The subsequent admission of 'errors' and the attempt to correct them was a compromise (the first of many) between the peasantry (or sections thereof) and the VWP leadership.

Chapter IV shows that the liberation of the peasantry in economic and political terms, and the ending of war and instability, produced a rise in agricultural output and productivity. The rise was unmatched in terms of growth rates and--in the case of rice--even in levels, in subsequent years. One consequence, given the acute shortages of consumer goods, was a 'scissors crisis' with implications for State procurement of agricultural output. In the absence of unlimited aid, resources from agriculture were necessary to finance the industrialisation strategy. Peasant marketing autonomy is shown to have posed problems for the State.

The VWP continued the momentum of agrarian change by developing Mutual Aid Teams. In 1958-9 it launched collectivisation based on three main arguments: that cooperatives exhibited increasing returns to scale and were an inevitable advance over peasant family production; that cooperatives limited rural differentiation which, if unchecked, would lead to capitalist agriculture; and, that only through cooperatives could the State exert control to 'tax' agriculture sufficiently to finance industrialisation. Additional factors included the influence of China and the leadership ideology--cooperatives were inherently part of the 'transition to socialism.'

The 'evidence' for the first two arguments is shown to be, at best, unconvincing, while the third relates to the need for control of agricultural surplus and hence of production. The development of cooperatives, peasant attitudes towards them, and the early problems they faced are also considered in Chapter IV. Even in these 'good' years two crucial problems can be identified: the managerial shortcomings of cooperatives and the limited supply of consumer and agricultural producer goods.

Chapter V discusses the problems faced by cooperatives in the years before the major impact of US bombing and the war. It shows that the VWP leadership had neither the strength to ensure effective collectivisation by force (a la Stalin) or by conviction, while the peasantry had insufficient strength to oppose and prevent the policy altogether, or to openly manipulate cooperatives in its favour. However, both practices occurred in certain places. The compromise is illustrated by the leadership's vacillation over material incentives, the divergence between 'form' and 'reality,' and the peasant tendency to devote resources to family production.

The limited availability of incentive goods required the use of other policies by the leadership, some of which are discussed. One of these was 'improved' management as a partial substitute for both technical change and material incentives. By the mid-1960s this had become essentially a call for more 'socialist' cooperatives and a repetition of the arguments advanced in the late 1960s. The slow growth of agriculture and the fall in food production per capita were the result of the problems of cooperatives and the constraints on the family sector, one of which was the short supply of consumer goods. The slow growth in turn made the mobilisation of food for urban areas and of resources for accumulation more difficult.

Chapter VI touches on problems of managing a war economy and its effects on agriculture. It shows that decentralisation had a number of impacts on agriculture. The relocation of some industrial production eased supply shortages while the war-related demand on resources worsened them. The war increased local autonomy and reduced State control over agriculture and the cooperatives, many of which continued to have a purely nominal existence, while others developed more as a reflection of members' interests than the State's. The halt in the US bombing after 1968 saw a reappraisal of the standard arguments. Policy responses continued to fluctuate between 'improved' management of cooperatives as a partial substitute for technical change, the need to curtail consumption to maximise accumulation, as well as the use of material incentives.

The leadership was continually shuffling the same cards in an attempt to find a combination which would 'solve' the problems faced by agriculture and the economy. However, the fundamental problem remained an absence of conditions sufficient for cooperatives to form part of this solution.

CHAPTER I

A Background to the DRV

The Physical Background

The territory of the DRV consists of 165,000 square kilometres. Its population in 1954 was 13.4 million and in 1975, 24.6 million. The vast majority of the population is concentrated in the Red River delta which is approximately 15,000 square kilometres. It amounts to only 10 percent of the land but contains over 75 percent of the population.[1] The population density in this area--over 1,000 people per square kilometre in certain places-- is among the highest in the world. Thai Binh, the major rice producing province, had a density of 1,054 per square kilometre in 1976. The remainder of the DRV consists of smaller deltas and plains along the coast and a large hilly and mountainous hinterland. The midland areas are considerably less densely populated and the highlands are very thinly populated (as low as ten people per square kilometre) and are inhabited largely by minority races.

The total cultivated area in the DRV is around 2.4 million hectares, over 75 percent of which is concentrated in the Red River delta. There are two climatic seasons: the cold, dry season from October until April and the warm, rainy season from May to September. Rainfall is irregular, however, and even the wet season may have periods of drought. The extreme violence of storms leads to floods and water-logging. The Red River itself causes extensive flooding as its level and flow varies enormously. It provides water for sowing, transplanting, and growth in addition to alluvial silt (130,000 tons per year). Continual maintenance and extension of the system of dykes is thus vital. Despite the efforts of the villagers and colonial government, serious breaches occurred in about one out of every two years from 1900-1930.

Given an adequate and controlled supply of water, the tropical climate permits year-round cultivation, especially of rice. Traditionally two crops are grown: the 5th month crop, sown in October-November and harvested in May-June, and the 10th month crop, sown in May-June and harvested in October-November. Two other crops--autumn and spring (the latter more successfully)--were developed during the 1960s. Rice occupies between 70 and 80 percent of the total cultivated acreage with subsidiary crops of maize, manioc, and sweet potatoes occupying between 10 and 20 percent. The remainder are vegetables, beans, and a variety of industrial crops.[2] The majority of the population is thus involved in the cultivation of rice in the Red River delta, although considerable effort has been devoted to transferring population to the highland areas and to encouraging crop diversification.

Administratively the DRV was divided into provinces, districts, and village-communes.[3] The exceptions were the Thai-Meo autonomous region and the two major cities of Hanoi and Haiphong. Villages could be extremely large and were subdivided into a number of hamlets.

The Impact of Colonialism

The basic unit of rural organisation in Vietnam for centuries has been the village-commune (xã). The commune enjoyed a substantial amount of autonomy and, to a degree, shielded its inhabitants from the intrusions of the State,[4] although by 'imprisoning' the peasants behind the 'bamboo hedge' it also put them at the disposal of the State and wealthy notables.[5] The commune was ad- ministered by a Council chosen from the oldest male inhabitants. Originally age was the sole criterion for Council membership but as a process of rural differentiation developed, wealth began to play a role, as did the State bureaucracy.[6]

The Council was responsible for the periodic reallocation of communal land[7] which originally comprised the majority of the village land. State exactions were levied on the commune with the Council determining individual contributions. Landed property consisted of holdings given to members of the nobility and supporters of the emperor. While nominally remaining the emperor's property, this land was effectively passed on as inheritance. During the period from the sixteenth to the eighteenth centuries, a growing landlord class consolidated its holdings and encroached upon the communal land by way of its position on the village council.[8] These tendencies, along with dynastic and demographic crises, led to periodic peasant uprisings which increased during the latter part of the pre-colonial period.[9]

The imperial administration intervened on occasion, especially at the beginning of a new dynasty, to reverse the process of privatisation of land.[10] However, as population density increased and agricultural production failed to keep pace, the potential gains from private property increased and differentiation developed.[11] Major technical developments did not occur in agriculture and increases in production resulted more from replication of the commune on new land than from improvements on existing land.[12] The development of markets was limited by conditions of geography (i.e. the existence of transport difficulties in a delta) and by the similarity of output in the closely packed communes.

French involvement in Vietnam began in the early nineteenth century and occupation in 1858-59. Full control was achieved and the country divided into three administrative regions--Tonkin, Annam, and Cochinchina--in 1884.[13] The DRV comprised Tonkin and the northern part of Annam. The French presence had two opposing effects on the village-commune. In the first place, it accentuated problems arising from the further penetration of a cash economy. Then, by largely preserving the existing institutional organisation, it contained the increasing class tensions engendered by the impact of the cash economy.

The French increased exactions on the peasantry to pay for the expensive colonial administration and the existing structure was a convenient means for meeting these expenses. The richer rural classes took advantage of the potential offered and acted as effective agents of the French. With the partial counter of the imperial State now removed, they increased their own exactions on the peasantry.

In short, the French can be said to have strengthened some aspects of the traditional hierarchical structure to the detriment of the majority of Vietnamese, while allowing a new indigenous minority to share in the returns as long as they remained obedient and necessarily insensitive to popular grievances.[14]

The primary aim of the French in Vietnam was to develop exports of tropical agricultural products and certain minerals. Infrastructural developments were related to this extraction of resources and commodities rather than to internal developments.

An exaggeration of the means of communication occurred before a corresponding development of the means of production.[15]

The labour force for these enterprises was obtained through various forms of 'recruitment' in the northern villages. Very little industry was developed since Indochina was considered a market for French products rather than a competitor.[16] It was hardly an important market given the extremely limited level of internal demand and the absence of efforts to develop that demand. It became more important as a market during the world recession of the 1930s, though given the falls in commodity prices and hence plantation wages and peasant income, demand also fell during this period.

The main exports during the colonial period were rice and the newly introduced rubber. Although rice production in the whole of Indochina rose by 47%--from 4.3 to 6.3 million tons between 1900 and 1937--the population rose by 80%[17] and per capita consumption is estimated to have fallen from 252kg to 182kg.[18] The annual rice ration in Tonkin was even lower at 136kg per capita. A minimum consumption standard is normally cited as 240kg of paddy per capita per year equal to 156kg of rice.[19] Much of the extra output was thus exported. The limited increase in the face of a doubling of the rice lands in Cochinchina indicates a slow technical development of production methods. The mean yield per hectare hardly rose during eighty years of colonial rule and by the late 1930s was, at 1130kg, the lowest of any major rice-producing country.[20] For Tonkin alone the yield was 1268kg in 1942-43.[21] For comparison the yield in China was 2500kg per hectare; in Burma 1400kg; in Thailand 1520kg; and in Japan 3826kg.[22] Ishikawa has argued that a yield of 1300kg per hectare was the basic yield in the region for paddy cultivation,

> on the same land on a continuous crop basis even without...basic investments, only if adequate sunshine, temperature and water supplies are available.[23]

Unfortunately detailed information on Vietnamese agriculture during the colonial period is extremely limited. The work of the two major French geographers, Yves Henry and Pierre Gourou,[24] is still used by the Vietnamese themselves. Such information as is available often relates to all of Indochina rather than to the Tonkin delta alone, which is the significant agricultural area of the DRV.

There is, therefore, no comprehensive data on yields, the extent of double-cropping and irrigation, or the level of agricultural inputs. The increase in double-cropping in the late 1950s suggests that a capacity for improvement within colonial agriculture existed. The low yield in colonial Tonkin in comparison with other countries also indicates an unexploited potential, even with existing inputs. This may however have been exhausted by the late 1950s. The average yield in the years 1957-59 in the DRV was 2040kg per hectare.[25]

It has been argued in the case of China, that by the late 1950s it had come to the end of many of the "pre-modern sources of increased grain production."[26] As in pre-colonial Vietnam one of these sources was the extensive margin. In terms of yields the DRV may have had greater gains to make than China, given its relative backwardness in the colonial period.

Agricultural techniques during colonial and pre-colonial times relied almost solely on labour.[27] Animals were relatively little used in the densely populated regions because they competed for the meagre food supply. There was virtually no machinery, although given the existence of labour shortages at certain times--a constraint on the extension of double-or treble-cropping--some mechanisation would have been advantageous. The labour shortage coupled with a lack of draft animals was seen as a major constraint on the extension of double-cropping in China in the 1950s.[28] Such few tractors as had existed (two hundred at most for all Indochina) were nearly all lost during the war years.[29] Fertiliser supplies consisted only of sources available to the peasant at no cost, and chemical fertiliser was virtually non-existent. Imports in the last four pre-war years averaged only 11,000 tons for all Indochina compared to an estimated requirement of over 200,000 tons.[30]

The irrigation and water-control network required continual maintenance. Much of the effort expended in the late 1950s involved restoration of the already existing system rather than building completely new works. On the basis of later evidence, it could cautiously be asserted that both the potential for and the returns to substantial new irrigation investment were limited.[31] In the case of China, again it has been claimed that "water control construction had long since reached a point of sharply diminishing returns."[32]

It seems possible to conclude tentatively that the output and productivity of agriculture in colonial Vietnam could be raised using traditional inputs, but that the returns to these would rapidly diminish and further significant increases would require increased modern inputs.[33] Furthermore as was argued in the introduction, there is no clear evidence that collectivisation is the most appropriate organisational form for maximising the yield from traditional inputs.[34]

The exhaustion of the extensive margin in Cochinchina, partly by French land-takeovers and allocations, removed one important release for the pressure of population on land.[35] Table 1.1 gives data for landownership in Tonkin during the latter part of the colonial period.[36] In addition an estimated 36% of rural families were tenants and sharecroppers and 20% of the land was at least nominally communal. As Table 1.1 shows, the majority of owned plots were extremely small and it is estimated that poorer peasants owned only 50% of the land they farmed.[37] In terms of the later agrarian reform data and classifications, 'middle' peasants in 1945 had an average of 1680 square metres per capita[38] or approximately 0.75 hectares per family. In addition many nominally owned plots were effectively mortgaged to money-lenders on whom poor peasants were often dependent for consumption loans. In some places, larger landowners apparently falsified registration documents to conceal property relationships. Share-cropping became more common during this period with rental being, on average, 50% of the crop.

Table 1.1: Landownership in Tonkin in 1930

Size of Plot	Number of Landowners	% of Total
under 0.36 ha.	594,000	61.6
0.36 to 1.8 ha.	288,000	29.9
1.8 to 3.6 ha.	60,000	6.2
3.6 to 18.0 ha.	21,000	2.2
18.0 to 36.0 ha.	800	0.1
over 36.0 ha.	200	---
	964,000	100.0

The communal lands contributed to agricultural stagnation or 'involution' in a number of ways. They inhibited migration, at least on a permanent basis, as peasants feared they might forfeit their rights. As the likelihood of obtaining land and similar rights elsewhere diminished, this factor increased in importance. They inhibited improvements in the land since the possibility always existed that it would be taken away by others, particularly those dominant within the village.[39] This factor reappeared later with cooperatives. Moreover, the pressure of increasing population raised the value of communal land and hence created disputes regarding its use and resulted in a diminition of individual shares.

The population density, attendant land tenure

systems, backward agrarian techniques, limited economic development, and high State exactions contributed to further rural impoverishment. Given the opportunities for letting and usury, landlords had neither need nor incentive to invest in productive improvements. As argued by Chayanov,[40] in a densely populated area of peasant farming, the rent of land will exceed its marginal product and peasants with land in excess of their minimum subsistence needs will find it in their interests to rent it out rather than work it themselves. There were also limited opportunities for agricultural accumulation because of the absence of an adequate financial infrastructure and the illiquid nature of rural assets, as well as the absence of capital inputs. Agricultural improvements required either capital input or mass labour mobilisation. The opportunities for investment outside of agriculture were also limited by the size of the market and French monopolies in such industry as existed. The constant need for rural credit resulted in peasant families lending any surplus cash they happened to accumulate.[41]

The structure of the colonial economy was such that the retarding forces, already present in part, were exaggerated and no favourable dynamic for general agricultural development and modernisation was established.[42] Although regionally selective agricultural development occurred, notably in parts of Cochinchina, State exactions increased the pressure on the peasantry above the pre-colonial level.

> After rent and usury, taxes are the third yoke on the neck of the peasant.[43]

The colonial administration had five major sources of taxation: two direct--the land and head taxes, and three indirect--on alcohol, opium, and salt. It has been estimated that the indirect taxes accounted for more than 50% of the total. The monopolies by way of which the indirect taxes were levied were particularly hated for they

> touched the indigenous population in three aspects of its physical existence: a vice, the opium; a habit, the alcohol made of rice; and a necessity, the salt.[44]

Tax rates tended to increase during the colonial period and estimates of the proportion of peasant income absorbed by taxes vary from 9% to over 25%.[45] Their collection was also permeated with corruption. It has been estimated that the repay- ment of debt absorbed some 10% of peasant in- come.[46] In addition to the 'three yokes,' surplus extraction also occurred through trade[47] and corvée labour. Since it was redeemable, corvée labour became yet another possible source of enrichment for administrators while it increased the monetary burden on the peasantry.

By the 1930s, the economic conditions of the mass of the peasantry could be characterized as follows:

> In general, the conditions under which the masses live are excessively miserable. The cultivator lives entirely on what he can get from the land and he can satisfy only his most immediate needs. He is even faced with the problem of hunger if a year's harvest is less satisfactory or if an unexpected event like sickness or accident causes him unforeseen expenses.[48]

This brief outline has not touched on the issues of 'nationalism' and 'Vietnamese consciousness' during the colonial period. These issues have been discussed extensively elsewhere[49] and they undoubtedly made important contributions to the development of patriotic resistance against the French. However, the extent to which these factors influenced the peasantry is not clear. The peasants would support <u>any</u> group which promised land, a lifting of the 'three yokes,' and, by the early 1950s, peace and the prospect of development.

<u>An Overview of the Development of the Economy and Industry, 1954-75</u>

This summary of the key aspects of economic and industrial development is necessarily brief for two reasons. There is an extremely limited amount of data on these aspects, particularly for the period 1965-75. Figures 3 to 7 present data on different aspects of economic and industrial performance compiled from a number of sources, and the breaks in the series are clearly evident. The data for these figures and some of the data used in the summary are presented in Appendix Table XI. The second reason for the brief treatment of industry is that this work focuses on agriculture and its history during the period. An economic history of industry in the DRV and indeed of the whole economy remains to be written. It should be added that frequent reference is made to industry throughout the work. Theoretical aspects of the relationship between agriculture and industry were discussed in the introduction.

Industrial production in the DRV in 1954 represented only 1.5% of total production.[50] Modern industry consisted of only seven enterprises. During the latter part of the colonial period industry had represented some 10% of total production,[51] but much of this had been destroyed by the war. The French also dismantled machinery when they finally withdrew. Prior to partition some 200,000 to 250,000 tons of paddy rice were sent from the Mekong delta in the south to North Vietnam.[52] By 1955, 90% of imports and 91% of exports were with 'socialist' countries and in 1957, 89% and 66% respectively.[53] The economy faced problems because of the very low industrial output, the virtual cessation of trade with the south, and the shift from trading links with France to the Soviet bloc.

The cut-off of French trade caused serious shortages of consumer goods and price rises in the period up to 1958. As a result, during the reconstruction phase (1954-57) considerable attention was given to light industry which accounted for 75% of industrial output and grew more rapidly than heavy industry. Imports of consumer goods were 55% of all imports in 1955.[54]

By 1957 industrial output had reportedly attained 76% of the pre-war (1939) level.[55] The state sector accounted for 28% of industrial production.[56] The country then embarked on industrialisation centered around heavy industry. Both the Three-Year Plan (1958-60) and the Five-Year

Plan (1961-65) accorded clear priority to the development of heavy industry--67% of industrial investment in the former and 79% in the latter. Industry was to grow much faster than agriculture and handicrafts, and heavy industry was to grow faster than light industry.[57]

The targets for the Five-Year Plan were lower than those for the Three-Year Plan as a result of experience as well as of the increasing constraint posed by agricultural development. The production of consumer goods reportedly grew at 16% per annum from 1958 to 1960 and 12% per annum from 1961 to 1964. Handicrafts, later admitted to have been neglected, grew more slowly (10% in 1958-60 and 6% in 1961-64).[58]

The low levels of income and domestic surplus meant a heavy reliance on foreign assistance during the two plans.

> In our Three-Year Plan, the construction of the major part of medium and heavy industry is to be realised by assistance from fraternal countries.[59]

One estimate is that foreign aid provided 62% of financial investment requirements for the Three-Year Plan (89% in grants, 11% in loans) and 34% for the Five-Year Plan (all in loans).[60] Foreign aid reportedly constituted 40% of government budget receipts in 1955, 41% in 1956, 38% in 1957, 26% in 1958, and 18% in 1960.[61] In 1960 accumulation was 59% of government expenditure. Foreign assistance was dependent on the fluctuations of the Chinese economy, international considerations --including the Sino-Soviet split--and the later war with the United States.

Domestic resources for accumulation for the plans came entirely from the budget. Government revenue during the reconstruction period came primarily from taxes, including agriculture.[62] Later it came overwhelmingly from trade via the State enterprises.

> The exchange between industrial and agricultural products helps realise accumulation through home trade and

> especially foreign trade, which makes it possible to import equipment, machines and raw materials for industry.[63]

In 1960 receipts 'collected from sales of goods' constituted 76% of the total.[64] They remained over 70% in the subsequent period.

The percentage of the workforce in industry was 7% in 1960, 8.3% in 1965, and 10.6% in 1975. The absolute agricultural workforce which was 5.9 million in 1956, fell to 4.9 million in 1970 and then rose to 5.8 million in 1975.[65]

The rate of growth of agriculture and, more important, its marketed surplus were key determinants of accumulation and the industrial growth rate.

> Every year an increasing amount of agricultural surplus must be realised in order to supply the expanding non-agricultural sector and to pay for increasingly necessary imports of equipment. Thus, it is not sufficient to obtain an overall increase in the volume of agricultural production: in addition, the State must have absolute control of the commercial disposal of the surplus, so as to avoid its loss through an overly rapid growth of consumption.[66]

Therefore, one constraint was that imposed on industry by agriculture. This can be seen in both the trend and fluctuations in the value of industrial and agricultural output.

Figure 2 shows dramatic fluctuations in the industrial growth rate which mirror, with a lag of one year, those fluctuations in the growth rate of agriculture. If industry were unconstrained by agriculture a much more stable growth rate would be expected.[67] In addition, Figure 2 clearly shows the impact of the two major periods of US bombing--1965-68 and 1970-72--on industrial production. However, another constraint operated in the other direction. In order to raise agricultural marketings the State had to be able to provide goods for the peasantry. The economic plan was based on heavy industry which received the major share of accumulation, even after 1965. In the period 1965-68 Group A received 82% of industrial investment while in 1969-71 it received 76%.[68] Heavy industry thus grew faster than light industry and the share of the latter fell from 73% in 1957-59 to 59% in 1963-65. It continued to fall during the war years despite the apparent increased emphasis on light and local industry.[69]

Local industry corresponds closely with light industry over the whole period in terms of overall growth rates. The development during the mid-1960s illustrates the localisation of a certain amount of heavy industrial production. The share of local industry fell to 46% in 1965, rising to 56% in 1973. Handicraft production grew the least of all the various industry and handicraft sub-sectors and was virtually stagnant during the 1960s; its share fell to one quarter of the total in the 1970s.

The handicraft sector was a major supplier to the peasantry and its performance was not compensated for by a sufficiently rapid expansion of industrial consumer goods production. Its performance may have been demand constrained whereas light industry's slow growth was partly deliberate (i.e. an initial emphasis on heavy industry) and partly due to slow growth in other sectors. In per capita terms, the supply of consumer goods peaked in 1964-65 then fell during the war before rising again in the 1970s. In 1973 it was only 50% above the 1960 level and barely above the 1964-65 level. The increase in the share of consumer goods in total imports from 13% in 1960 to 32% in 1974 may have been an attempt to alleviate this constraint. However, it is not known to what extent the import of consumer goods included food.

The emphasis on heavy industry and the neglect of handicrafts was a constraint on the development of agriculture.

> Many provincial engineering enterprises have devoted 80% of their production to other industrial branches instead of working for agriculture.

> This tendential decrease of handicrafts is but the result of an under-estimation of the role of local industry, and of handicrafts in the first stage of industrialisation, particularly as regards agricultural development.[70]

This constraint was later admitted more explicitly by the DRV leadership:

> For the past twenty years, we have had a virtual absence of industrial machinery for agriculture, and also not enough light industry, so our cities didn't have much to sell to the countryside. As a result, the peasants were not keen to sell to the city folk.[71]

In addition, there was frequent criticism of the poor quality, irregular supply, and in appropriateness of the goods available, regardless of whether they were adequately maintained or used.

> In many areas, between 30% and 60% of agricultural tools and machines are out of order due to the fact that they are of poor quality and unsatisfactorily maintained and used.[72]

The availability of industrially produced consumer goods to the peasantry would, in the face of the limited supply, depend on proximity to urban areas. It would be expected that a major portion of these consumer goods would not penetrate into large areas of the country, hence the importance of handicraft production for the mass of the peasantry.

The major change in employment was a small rise in the share of industry. In absolute numbers those employed in industry more than doubled. Frequent criticism was made of labour productivity in the industrial sector.[73] The population of those of working-age increased by almost 50% between 1960 and 1975. The major increases in shares of employment were in construction--from 1.7% in 1960 to 7.5% in 1975, and in 'non-material production'--from 3.4% to 10.7%. The bulk of the latter occurred in health and

Figure 1: Rice and Total Staple Food Production (million tons) 1954-75

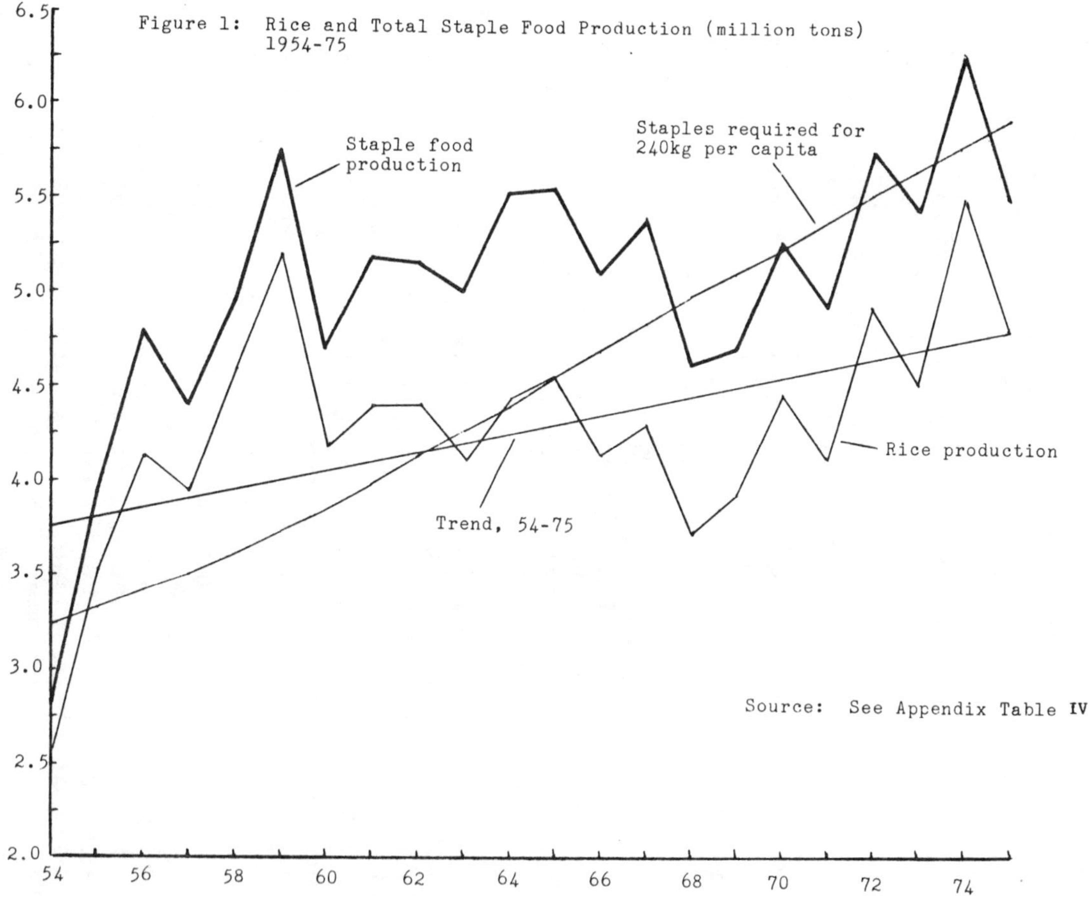

Source: See Appendix Table IV

Figure 2: Annual Changes in Gross Value of Industrial and Agricultural Output, 1956-75

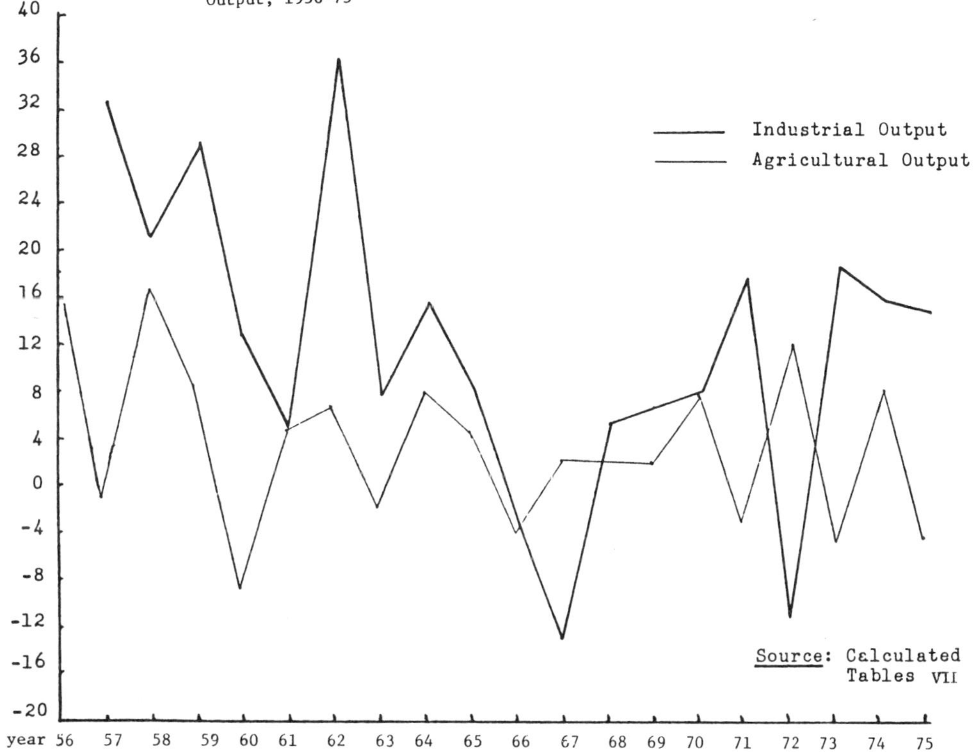

Source: Calculated from Appendix Tables VII and XI.

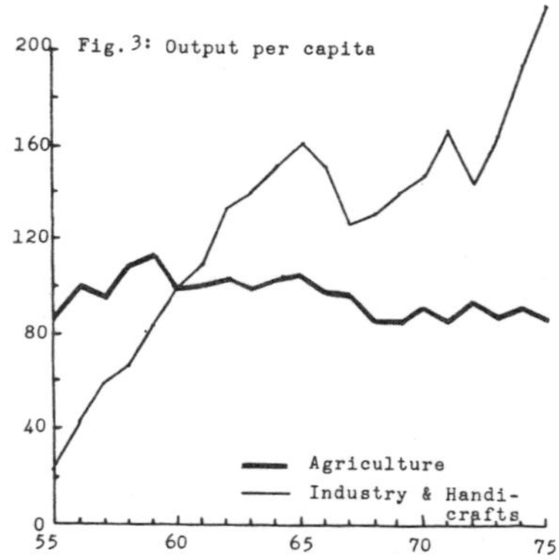

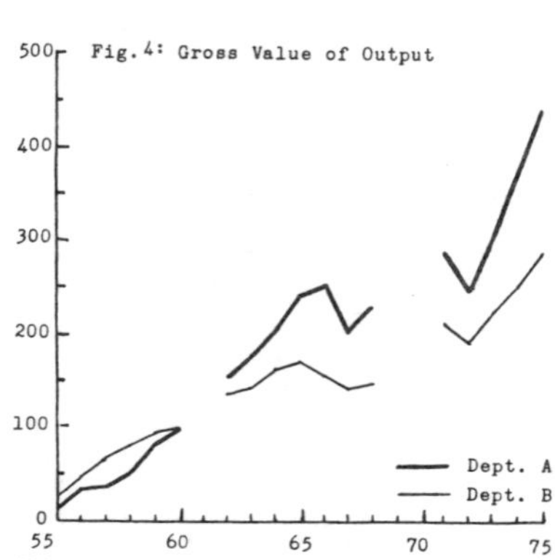

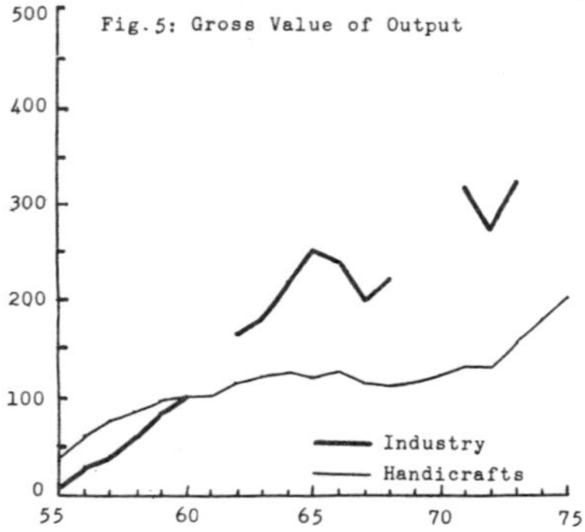

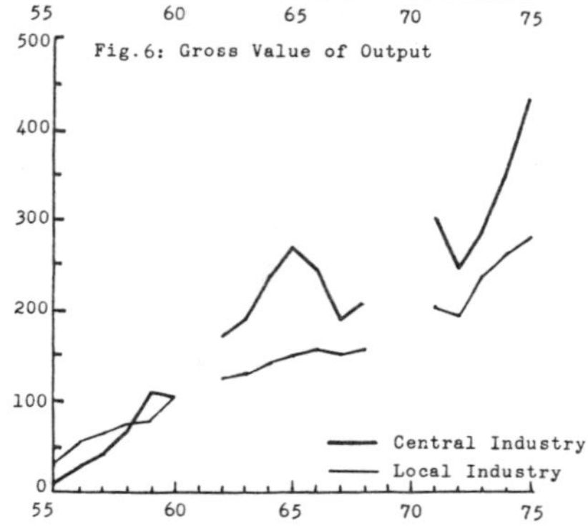

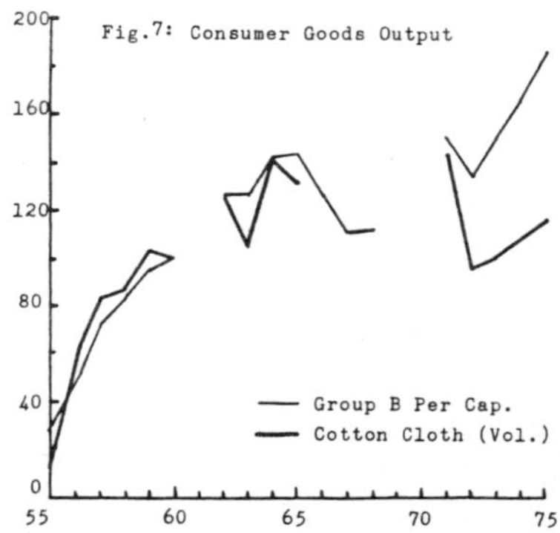

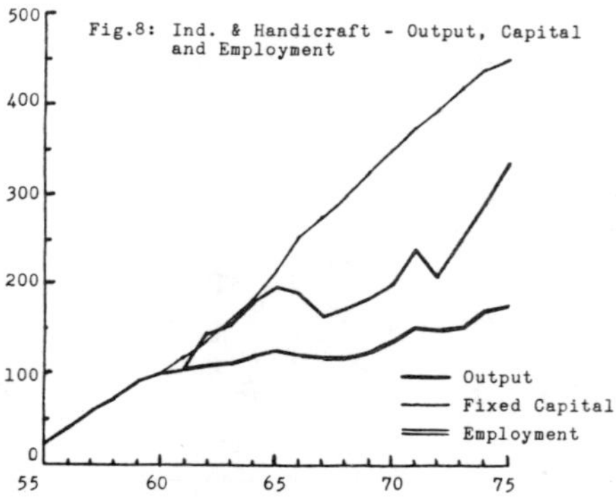

Note: All indexes at constant prices. Source: See Appendix Tables IV, VII, XI.

education which reflects the enormous improvements in these areas during the period. Consequently a fall occurred in the share of employment in agriculture and forestry--from 83% in 1960 to 65% in 1975, and apparently from 77% to 53% for agriculture alone.

Employment growth in non-agricultural sectors provided an alternative source of income for the peasantry both directly from employment and indirectly from demand. The agricultural workforce fell more rapidly as a percentage of the rural workforce--from 88% in 1960 to 62% in 1975, indicating that much of the alternative employment growth occurred in the rural sector. However, the growth in unproductive activities took resources that might otherwise have been used for accumulation in agriculture or light industry.

In 1975 the DRV was still overwhelmingly an agrarian economy. Agriculture accounted for 40% of national output as opposed to 43% in 1960. Industry's share had risen in 1965 and then stagnated, largely due to the consequences of the war which had also resulted in a fall in the share of trade from 18% in 1965 to 13% in 1974.[74]

This outline of the structure and evolution of the DRV economy provides a background for the consideration of agriculture. Agriculture constrained industry and industry constrained agriculture, given the absence of unlimited aid and the development strategy chosen; the extreme backwardness of the economy; the very small initial industrial sector; and the limited agricultural surplus.

Political Background

The Indochinese Communist Party (ICP) was founded in 1930 on Comintern orthodoxy,[75] and in organisational terms has always been 'Stalinist.' Its early development was influenced by a number of factors;[76] its leading cadres were in prison following the 1930-31 uprisings,[77] and it adopted the changing Comintern line, although with a lag. For example, the ICP Plenum of July 1936 adopted a united front policy following the line of the 7th Comintern Congress in July and August 1935 and abandoned its earlier 'class vs. class' line which had included criticism of Ho Chi Minh and his nationalism.[78] In addition, many of the ICP's leading cadres gained experience in China which influenced their attitudes and policies. In these early years the ICP's prime ideological focus was on the working class and alliances with other urban nationalist forces. Its policy with respect to agriculture was to obtain peasant support by promising 'anti-feudal' measures.

The remaining major influence on the ICP was Ho Chi Minh[79] who had extensive international experience. He had travelled and worked in the West and had been a founding member of the French Communist Party. He had attended many important conferences in the Soviet Union in the 1920s when he argued a pro-peasant line.[80] As Comintern agent he had worked all over Southeast Asia and had gained important experience in China during the late 1930s and 1940s.[81] These factors account partly for the opportunist nationalist line often pursued by the ICP/VWP and the absence of anti-intellectualism which appeared in China.

Ho returned to Vietnam, reportedly for the first time in thirty years, to preside over the seminal 8th Plenum of the 1st ICP Central Committee (CC) in May 1941. This followed the abortive Bac Son uprising in November-December 1940 which cost the ICP more members and supporters.[82] This meeting laid down a broad 'united front' line and led to the founding of the Viet Minh.[83] The slogan, 'land to the tiller,' was effectively dropped and agrarian reform policy was limited to the redistribution of communal land and rent reduction. The meeting recognised the need for a broad alliance of all 'patriotic' rural and urban groups, as well as the need for secure bases and extensive organisation and propaganda in rural areas.[84] The ICP was very weak in the 1930s, partly because of the two abortive uprisings, but also because of the success of the colonial police authorities.

The 8th Plenum blended together many of the elements referred to above and the line it established remained dominant for the following ten years. By the end of World War II, the Viet Minh had strengthened its position and influence but it was neither the only nor the dominant 'nationalist' grouping. However, it seized the opportunity presented by a power vacuum and presented itself to the allied forces as a new administration. Influenced by a number of factors, Ho Chi Minh initially pursued a policy of negotiation with the French. There was the possibility of US support, although this faded after the death of Roosevelt. There was also the pragmatic desire to get the Chinese Nationalist forces withdrawn. In addition, the Soviet Union urged caution since it hoped for a Communist victory in France, and the French Communist Party wanted to retain Indochina.[85]

As these factors changed and the prospects for any successful negotiation receded, armed resistance became more likely. The Viet Minh, however, retained the nationalist card and stressed an alliance of all groups whose objective was independence, while the ICP officially 'disbanded' itself. The broad alliance was both economically and politically necessary since the DRV administration and army relied on the output of surplus producers.[86]

Stronger Soviet support, as well as the Chinese Communist victory and recognition of the DRV in 1950, brought about the official re-emergence of the ICP as the Vietnam Workers' party (VWP) and the decision to follow Chinese policy. The internal balance also changed in the early 1950s (see Chapter II), and there were socio-economic reasons as well as international political considerations for abandoning the 'united front.' This led to the monopolisation of power through the destruction of potential or actual political opponents and to the implementation of a major agrarian reform which went beyond policies so far implemented. For reasons discussed below (Chapter II), these changes caused major internal economic and political difficulties which necessitated corrective action. The timing of this action (admission of 'errors' and Rectification) was partly determined by developments in the USSR and China in 1956 which provided an ideological rationale for it.

Reliance on Soviet and Chinese aid and cadres

in the 1950s contributed to the following of an orthodox 'Stalinist' industrialisation policy. This involved a focus on heavy industry and a squeeze on agriculture to which collectivisation was intended to contribute. The slower pace at which cooperatives were developed in the DRV, compared with those in the USSR or China, was a result of the fundamental problems the cooperatives encountered as well as the inability of the VWP to pursue the Stalinist approach of force. There were a number of reasons for this: VWP cadres were not all 'communists'; the VWP itself faced problems because of the unification struggle; and the consequences of a 'peasant strike' would be disastrous. In practice, the VWP tended to be more sympathetic to the peasantry (as a whole) than the political line suggested, although this was also partly due to economic and political expediency.

Far from following a distinct or independent 'socialist' development path, the DRV has consistently adopted a policy package based heavily on various aspects of Soviet and Chinese policy and practice. From its formation as the ICP, the VWP has had an urban orientation, although under Chinese influence this shifted somewhat towards the peasantry. Both its relatively limited rural penetration and the divergence of interests between the VWP leadership and the peasantry have contributed to the problems encountered with collectivisation. The final notable characteristic of the VWP leadership is its remarkable stability. It has experienced none of the purges and dramatic changes characteristic of other communist parties.

CHAPTER II

From Colonialism to Partition: Rural Policy Developments, 1939-53

Introduction

French colonial rule in Vietnam effectively ended with the collapse of France in Europe in 1940, although the French did not leave until 1954 after the Geneva agreements and cessation of hostilities. During the fourteen-year interregnum, Vietnam was under Japanese suzerainty for five years and thereafter was fighting a war of independence.

The disruptions caused by the fighting and instability of these years worsened the living conditions of the peasantry as agricultural output did not rise and population continued to increase. Moreover, the Japanese increased the rate of exactions above the level reached during the colonial period, while from 1946 to 1954 there were two administrations (the DRV and the French) attempting to tax the peasantry.

The nationalist resistance against the French was led by the Viet Minh, within which the Indochinese Communist Party (ICP)--later the Vietnamese Workers' Party (VWP)--played a dominant role, having entered the political power vacuum left in mid-1945. This movement gained the cooperation and support of the peasantry by promising to fulfill its main demand--land. The absolute scarcity of land and the impossibility of fulfilling this demand without appropriating a large number of peasants with above average holdings caused major problems during the agrarian reforms. Unfortunately, Vietnam--especially that part which became the DRV--had very few large landowners, and those landlords who did exist did not have large amounts of land in absolute terms.

Implementation of distributionalist agrarian reforms was constrained during the resistance period by the maintenance of a 'united front' policy. The landlords and rich peasants incorporated within this front were important for their political influence and local authority, and also provided a large part of the surplus production on which the Viet Minh army and government depended. International considerations also dictated a moderate policy towards these groups. The agrarian policy of the DRV and VWP was initially limited to rent and interest reductions and distribution of land appropriated from the French and their supporters.

For both economic and political reasons, the balance between fulfilling the promise of land and appeasing the richer rural classes shifted in the early 1950s. The widening of the revenue base of the DRV by an increase in the area under its control, tax and monetary reforms, and assistance from China, all contributed to reduce its economic dependence on the richer strata. Pressure also increased from the poor and landless peasants whose aspirations had been raised by the education and literacy campaigns, and who had contributed to the war effort by providing recruits for the army, but had yet to receive sufficient land. The victory of the Chinese Communists and completion of their agrarian reform provided ideological strength for the argument for a major agrarian reform in the DRV. It also served to consolidate the monopoly of power by the VWP by removing potential political opponents.

While these forces were dictating the form and timing of agrarian reform, changes within the agrarian sector were increasing the likelihood of major problems. By 1953 over half of the land available for appropriation--if only those above rich peasants were to be appropriated--had been 'temporarily' reallocated. This had raised the landholdings of many peasants above the mean, while the majority of the poor and landless had insufficient holdings for basic subsistence. The middle peasants thus had little to gain in terms of land, from further distributionalist reform, but if the reform involved egalitarian pressures they had land to lose. They would not oppose 'anti-feudal' agrarian reform since it could strengthen their position by removing the rural class above them. The nature of the cross-section land distribution in 1953, in conjunction with the absolute shortage of land per capita, constituted a fundamental constraint on an egalitarian agrarian reform.

The Japanese Interlude

The Japanese intervention in Vietnam between 1940 and 1945 differed from the French because it never involved a direct relationship with the rural areas. Administration took place through Vichy French intermediaries. This resulted in a rate of extraction even higher than that carried out by the French alone prior to 1940. Food requisitions in Tonkin alone amounted to 130,000 tons in 1943, and 186,000 tons in 1944.[1]

Japanese aims in Indochina bore much superficial similarity to those of the French in that the area was an important source of food and raw materials and provided access to China from the south. In the longer term, even the methods of control and rule may have taken one of a number of direct or indirect forms of colonialism. However, from 1940 to 1945 the Japanese were content, for strategic reasons, to maintain a dominant military presence and use the existing French administration to fulfill their requirements. The result was the continuation of the immediate rural power struggle between the French and the independence forces, one of which was the ICP.

The French reliance on traditional village-commune leaders was one means of countering the growing insurgency, stabilising the village-communes and implementing the strict fiscal and security controls necessary to mobilise sufficient rice to fulfill both their own requirements and those of the Japanese.[2] The increasing hardship not only strengthened the independence forces but it also further discredited the existing village-commune leadership and accelerated the disintegration which had been developing during the previous period.

In the rural areas, the policy of the ICP and the Viet Minh liberation organisation--formed largely under its aegis during this period--was to confiscate the possessions of the French and Japanese imperialists and Vietnamese traitors, and to ensure the reduction of rents and interest rates as well as the redistribution of communal lands to the peasantry. This policy line had been laid down at the 8th Plenum of the 1st ICP CC held in May 1941, under the chairmanship of Ho Chi Minh, and remained the keystone of ICP policy for at least the subsequent ten years.[3]

The original policy line contained in the 1930 manifesto had been 'to requisition all lands belonging to the foreign and native landlords and the churches; and to hand them over to the middle and poor peasants.'[4] This was a nationalist policy designed to bring peasants into the struggle by way of anti-feudal policies. The 6th Plenum of the 1st ICP CC in 1939 had modified this in light of the situation at that time and Comintern policy as laid down at its 7th Congress in 1935: 'at this time the bourgeois democratic revolution only advocates confiscating lands belonging to landlords who are traitors to the nation.'[5] The 8th Plenum thus elaborated and consolidated this line, and removed all references to land redistribution with the exception of the communal lands.[6]

The French role in implementing the vicious taxation and control measures necessary to fulfill Japanese requirements increased the peasantry's hatred of them, and helped the reconstruction of the communist movement. The Japanese failure to construct their own administrative apparatus meant that when they took complete control in March 1945 (when the French were either interned or had fled), there was a void into which--at least in Tonkin--the Viet Minh could step as a national front, with US support. There was then a clear dichotomy between the independence forces led by the Viet Minh which included many landlords, and the Japanese enemy and its quislings. The Viet Minh was able rapidly to spread its influence and control in the rural areas, with an avowedly nationalist policy and reformist programme designed not to alienate the majority of landlords, and hence to ensure maximum support. At the time of the Japanese surrender in August 1945, the strength and authority in rural areas of the Viet Minh was greater than its military force may have implied.[7]

The Anti-Colonial Struggle, 1945-53

During the period of Japanese occupancy of Vietnam, the ICP and Viet Minh had accorded priority to achieving independence from both Japan and France. With the surrender of the Japanese and the return of French military forces, the priority of defeating the French remained. Only with changing circumstances around 1950 did policy shifts begin to emerge.

The USSR had neglected Indochina in 1945 and had taken a moderate line until 1946, hoping that the French Communist Party--which wanted to retain Indochina--would win power in France. Ho Chi Minh thus took the line of compromise, with independence in the future, and only gave up negotiations in December 1946. In 1947 the USSR gave verbal support to the Vietnamese 'national liberation war,' but it was 1948 and 1949 before it gave the effort much emphasis and full approval. Only after China recognised the DRV in January 1950 did the USSR do so.

The primacy of the liberation objective--the 'anti-colonial struggle'--inevitably meant that the 'anti-feudal' objective to which the ICP had committed itself at its founding in 1930 and which was embodied in the slogan 'land to the tiller,' would have to be put aside until conditions were more propitious. Thus, the agrarian programme was initially confined to confiscation of lands belonging to the imperialists and the Vietnamese traitors and their distribution to the poor peasants; redistribution of the communal lands; reduction of land rents and interest rates; and a moratorium on debts.

The immediate economic objective in 1945 was to increase the level of production to alleviate the famine conditions which had resulted from excessive taxation and forced extraction by the Japanese, who even used rice to make alcohol for fuel.[8] In addition, following the French return to Hanoi and the relocation of the DRV leadership to the mountainous Viet Bac area, it was necessary to construct a war economy with its concomitant taxation and mobilisation measures.

These objectives required the maximum support of all classes and policies were designed with this in mind. The result was that the DRV was able to survive the early years and build sufficient bases among the population to wage a protracted struggle. However, it was later questioned whether these policies were pursued for too long and, even within them, whether the anti-feudal struggle was accorded too little significance.

Truong Chinh argued as early as 1946 that insufficient attention was being paid to the implementation of certain policies such as rent and interest reduction, and to the development of agrarian reform policies which would erradicate pre-capitalist relations of production.[9] In particular, he argued that the 'stages' notion of revolution which involved moving from the anti-imperialist, to the anti-feudal, to the socialist phases, was incorrect and that simultaneous advance on all fronts was necessary. Such arguments may appear to contradict elements within the Viet Minh which wished to limit the struggle and not progress to agrarian reform involving the complete expropriation of the landlords; or they may appear to derive from a belief that more radical policies were required to maintain the struggle and lead to a more rapid victory.[10] A trade-off occurred between keeping up output and therefore being nice to the richer rural classes, and mobilising more support for the war, and hence helping the poorer middle peasants. Truong Chinh's comments derived as much from practical as from ideological arguments.

Limited agrarian policies were promulgated as early as September 1945 when taxation reforms were announced. These included the abolition of the head tax, excise taxes, a 20% reduction in land taxes, and the establishment of special taxes to finance the DRV government and resistance against the French.[11] Numerous other taxes were subsequently introduced until 1951 when the DRV taxation system was completely reformed and greatly simplified.

The earliest pronouncement on land tenure was the issue of an order in November 1945 for a 25%

reduction in land rents and the cancellation of all secondary rents, as well as all rent arrears owed by tenants before the revolution.[12] The first land confiscations occurred during the summer and autumn of 1945, but in all probability consisted of land previously abandoned by its former French owners.[13]

These policies do not appear to have been implemented during the next two years (at least) for a number of reasons: neglect by the leadership which was preoccupied with other events; the relatively limited area under permanent occupation by the DRV; and the danger of antagonising any section of the landlord class until the resistance had established secure bases within the geographical area and the population itself. In January 1948, the 2nd Plenum of the 1st ICP CC redefined the main lines of its agrarian policy emphasising the measures covered by the November 1945 order.[14] The policy was subsequently repeated in decrees of July and August 1949. Further DRV decrees of February and May 1950 dealt with the temporary distribution of the land of absentee landlords and the utilisation of presently uncultivated land.[15]

The emphasis of agrarian policy during this period from 1945 to 1950 was thus: the mitigation of peasant conditions via rent and interest reductions; progressive taxation and wage controls; a reversion to the traditional system of periodic redistribution of land owned collectively by the village-commune; and encouragement of increased production via utilisation of uncultivated land. Such a policy was required for a number of reasons already mentioned, namely, the need to limit the enemy to the colonialists and traitorous landlords in order to ensure maximum support for the resistance; the precarious position of the resistance struggle itself and the instability throughout the rural areas; the limited degree of class consciousness among the peasantry and the time required to raise this; and the relative paucity and inexperience of cadres who required time to be trained and accumulate the necessary experience before major agrarian reforms could be attempted.

These circumstances began to change in 1950, and by 1953 were considered propitious for the implementation of an agrarian reform programme which would destroy the economic and political base of the landlord class in the countryside. In the same way that the policy was called over-cautious and 'rightist' in the early resistance years, it was adjudged to have become 'leftist' during the agrarian reform campaigns as a result of a somewhat simplistic view of conditions in rural areas, particularly the changing class structure and the relative inexperience of cadres.

During the period 1950 to 1953 the fundamental line of government and VWP policy remained the same, namely, the need to maintain a broad united front to pursue the anti-imperialist independence struggle. Greater efforts were made to ensure the actual implementation of these policies and the formulation of more comprehensive policies began. In addition, the taxation system was restructured to put the nascent State's finances on a firmer footing and weaken the economic position of the landlords.

The changing national and international conditions contributed to this strengthening and radicalisation of policy and were accompanied by the official re-emergence of the ICP as the VWP in 1951. These new conditions included the final victory of the Chinese Communist Party in 1949 which secured the rear areas of the DRV and eased the supply situation. There was also a shift in the military situation in Vietnam in favour of the DRV resistance forces, although certain areas of the country had been under DRV control for some time by 1950.

Consequently, from 1950 the decrees relating to land rent reduction and provisional land redistribution were increasingly applied. The 'donation' of land to the peasants and State, and the reduction of interest payments and commutation of debts also increased.[16] A new, steeply graduated taxation system was introduced in a decree of July 1951 in order to reduce the net income of landlords and rich peasants. This tax replaced the existing range of taxes on land, trade, slaughtering, forestry, and many other activities, and consisted of a single contribution in grain, proportionate to the annual income plus an additional 1% tax paid to local authorities (although this latter element was subsequently reduced). The method of calculation was to divide the total family income by the number of members living on the land, and then to levy the relevant percentage per person. Initially the threshold was set at 60kg per head, though this was later raised to 70kg in July 1952, and to 80kg in August 1953. Those in the lowest tax band paid 5% in tax, while those in the highest ban--with rice income of over 1,816kg--paid 45%. Peasants who rented-in land paid 75% of the relevant amount, while those renting-out land paid 125%.[17]

There were numerous exemptions to the tax as well as incentive provisions to encourage cultivation of non-rice crops. Tax collection was hampered by the inadequacy of village records. The French proposed carrying out a major cadastral survey on a number of occasions, with the last attempt--which was never completed--beginning in the 1930s.[18] Neither the French nor the DRV considered the actual village-commune records reliable on the grounds that they had been adjusted and falsified by the village notables for a variety of reasons.

Major exhortations were made to complete the collection of taxes, although evasions and 'understatement' caused difficulties.[19] Both the increased revenue it generated and the wider tax base resulting from the increased area under DRV control contributed to an improvement in the budgetary situation. A by-product of the taxation system was that it produced a certain amount of estimation of landholdings and yields which would be useful in subsequent agrarian campaigns.

The real tax burden on the peasantry was increased by the need to pay both the French and the DRV in certain areas. An examination of this period shows that neither the French nor the DRV had permanent control of a substantial number of village-communes for any extended period of time. This makes it impossible to present reliable production figures for the period, and is one reason why consideration of the post-1945 achievements always tend to be compared with 1939 or earlier.[20] The absence of any coherent French policy towards the village-communes or agriculture during this period justifies the focus on DRV rural policy.[21]

The public relaunching of the ICP as the VWP in 1951 brought explicit recognition that the DRV, following the Chinese example, would proceed towards major agrarian reform, albeit at a slow pace. In a speech in 1951 to the Peasants' National Salvation Association, Truong Chinh repeated that the main tasks were the reduction of rents and interest rates; the implementation of regulations governing leasing of land; the distribution of French-owned land; the redistribution of communal land; and the utilisation of the land of absentee landlords.[22] He also emphasised that agrarian reform was necessary to guarantee the longer-term interests of the mass of the peasantry, and that land redistribution policies would be implemented. However, the development of these policies would depend on the availability of cadres and the situation in particular areas. In this speech Truong Chinh said the ultimate objective was the development of collective farming, but stated that this was still a long way off. There is a striking similarity to Chinese policy in this speech, although Truong Chinh was later to criticise a 'mechanical' application of Chinese policy.[23]

This speech can be seen as an explanation and justification for existing policies, as well as an outline of future objectives. It was also a criticism of those cadres, particularly in the southern regions, who were proceeding to the development of collective farms.[24] Truong Chinh's message was that, while policy should not be identical in all areas but take into account particular objective conditions, major agrarian reform was still in the planning stage, and collectivisation was definitely not on the immediate agenda. This speech may also be indicative of partial attempts to reassure landlord and rich peasant groupings that their positions were currently secure.

The balance between appeasing the richer rural classes and 'buying' poorer peasant support by the promise of land had shifted in 1951. The victory of the Chinese Communist Party had strengthened the VWP's position while the military situation had also changed. The VWP leadership thus decided to proceed to major agrarian reform. It was, however, another two years before the Rent Reduction and Land Reform campaigns were launched. The risk still remained that the early instigation of radical agrarian reform could increase support for the French among those groups which still retained considerable power, influence, and prestige. A key portion of output was also at risk.

The VWP leadership thus spent additional time laying the groundwork for implementation of the agrarian reforms before conditions were deemed 'ripe'. The agricultural taxation policy along with other financial problems absorbed a considerable amount of effort, in 1951 at least. A currency reform took place in mid-1951--one new dong for ten old[25]--and was coupled with a reduction of the amount of paper money in circulation in an attempt to reduce inflation.[26] Despite the issue of war bonds, valued according to the rice price and bearing interest of 3% over 5 years,[27] and exhortations for the peasantry to save (loan to the government) up to one-third of the harvest, the agricultural tax constituted the major part of government revenue during this period, providing 83% of the total in 1950-51.[28] Its relative importance diminished later in the 1950s as government revenue from new industrial enterprises increased.

In 1952 a major internal VWP rectification campaign was launched in order to remove from the VWP those elements which could pose as major opposition to later policies, and also to raise the ideological and educational level of cadres for the subsequent campaigns.[29] The rectification was directed against bureaucratisation, 'commandism, subjectivism, individualism, separation from the masses and reality, corruption and waste.'[30] It was also a response to the relatively rapid expansion of the VWP in the previous period when the expressed aim of DRV policy had been more placatory towards the landlord and rich peasant classes. It signalled the end of the 'united front' policy.

The crucial point with respect to the policy

changes was that it was now determined (and had been all along) that a distributionalist reform, consisting of the expropriation of landlords and the granting of usufruct or ownership rights to the peasantry, was the 'price' for peasant cooperation in the struggle against the French.[31] The peasants wanted land above all else. However, in the early years an all-out attack on the landlord class could not be carried out because it would disrupt the united front, and because the Viet Minh was comprised of many landlords and their sympathisers. At the economic level, the DRV was dependent on the landlords and rich peasants for the surplus food with which to support the army and administration.

One aspect of the policy debate at all levels of the resistance movements and the VWP, was the point at which mass peasant support might become more crucial than that of the landlords. Within this debate one view, illustrated by Truong Chinh in 1946, was that mass peasant cooperation was more important all along.[32] Others might have held the view that a major confrontation with landlords was not required; that what was needed was legislative action to limit rents, interest rates and the alienation of communal lands.

It is not clear to what extent alternative options, such as further legal action to protect tenants, confer inalienable usufruct rights, and control the credit market, were seriously considered at the higher levels of the VWP. However, the decision in 1951 to put agrarian reform firmly on the agenda and return more closely to the original 1930 policy line[33] which advocated confiscation of all landlord holdings and their distribution among the peasantry, necessitated a protracted internal struggle against those with opposing positions.

It could also be argued that the united front policy was becoming increasingly unnecessary by the early 1950s as the resistance forces gained strength. In addition, the monopolisation of power by the VWP required action against those who were both economically and politically dominant in local affairs. These objectives could be accomplished by the 'purge' and by agrarian reform.

Further strains within the nationalist coalition emerged as the richer classes bemoaned the cost of the resistance struggle and considered the alternative administration which was being assiduously groomed by the French and Americans. As a result of the education and literacy movements, some of the poorer villages had increased their political awareness and now resented bearing the physical burden of the struggle when there had been relatively little change in their socio-economic position.[34]

All the data relating to the agrarian reforms were derived either from official surveys of varying and frequently unspecified sizes, or from the aggregation of reports by Land Reform cadres and local authorities. The reports from local authorities are likely to be the most comprehensive, while all three forms are liable to bias of different kinds. Aggregation may exaggerate these biases or reduce their impact on the final results. The quality of the survey data also depended on the area(s) chosen, and on the ability and availability of statistical cadres. The most obvious bias is that data tended to reflect the VWP line which was current at the time of collection or publication.

The classification used in all the data is the one found in the 1953 guidelines, amended in 1955 (outlined in Chapter III). These guidelines are then hypothetically applied, retrospectively in most cases, to estimate the numbers who would have been classed in each category in any particular year. The land owned or used by each group was then estimated. Thus, data for land owned by (e.g.) 'landlords' in 1945 as opposed to 1949, refer to land owned by those who would have been classed as 'landlords' in 1945 as compared to land owned by those who would have been so classed in 1949. The individuals involved were <u>not</u> necessarily the same since the composition of the classes changed over time. This is shown for 'landlords' in Table 3.3, and would be expected for other classes as a result of the peasant family life-cycle, and of exogenous factors which had a differential impact on family fortunes.

Since most official data come from surveys

carried out post-Rectification in 1957 or 1958, the potential errors involved in estimating classifications and landholdings as far back as 1945 were considerable. Moreover, there were no reliable cadastral or population records to consult, so it was necessary to rely upon the evidence of the peasants.

Thus the degree of accuracy of the data is extremely limited because of the potential errors and biases involved. With respect to 'rich' peasants in particular, given their small weight in total population and large weight in land, enormous discrepancies can, and do, show up. Differences in data relating to population and families often have a simple explanation, namely, that family size is a function of social class, i.e. richer families have more members. An additional explanation for certain differences is the degree of severity with which the classification was applied in different villages and at different times (see Chapter III).

The data presented in the text must be considered with these points in mind. It involves hypothetically applying the classification to different points in time and the terms 'middle' peasant, 'rich' peasant, etc. refer to these groups at different points in time and <u>not</u> to the same individuals. The terms 'landlord', 'rich', 'middle', and 'poor' peasant are used in single quotation marks throughout when referring to those so classified or other official DRV usage. Middle peasant used here without quotation marks is defined unambiguously as a family with the ability to earn a per capita surplus (over basic food consumption requirements) in a particular year, assuming relative prices in that year to be given. Poor peasant accordingly refers to a family lacking the ability to produce a per capita surplus, while rich peasant refers to a family with the ability to earn a higher per capita surplus in the following year.

There appear to be no data which examine the presumably enormous changes in individual family circumstances during this period. Official data showing changes in the situation of families in subsequent years are discussed in Chapter IV, and are based on classifications fixed during Rent Reduction and/or Land Reform (subject to certain changes during Rectification). This fixed classification of individual families was retained, in principle, in later years. The impact of the inevitable changes wrought by the peasant family life-cycle and the agrarian reforms limited its usefulness in practice.

All data attempt to take, or retrospectively compile, a snap-shot of a constantly changing distribution. Given the conditions of the period, the high incidence of births and deaths, and the low life-expectancy, individual family circumstances could change considerably in a very short period of time.

Appendix Tables I and II summarise some of the available data relating to the overall impact of the agrarian reforms and illuminate many of the points made here. The data presented in the text are used to highlight the ambiguities of the classification guidelines themselves and the general characteristics of the structure of landownership and use in the DRV. They also illustrate certain obvious inconsistencies. The argument presented in Chapters II and III is not dependent on accurate time series data, but rather on the approximate cross-section land distribution in particular years. Arguments which do rely on the official data relating to the agrarian reforms must be treated with extreme caution given the above qualifications. This point also applies to subsequent periods.

The tables which follow survey the reported impact of land redistribution measures carried out prior to 1953. They show that while a certain amount of land was redistributed to the benefit of 'middle' and 'poor' peasants, the basic stratification of land ownership remained, with some 2% of the rural population owning almost 20% of the cultivated acreage. The effects of the rent and interest reductions and the taxation changes cannot be quantified but they would have partly mitigated the conditions for the mass of the peasantry, depending on the areas in which they lived. Some of the population were still subject to the rural policies and taxations of the French while certain areas would have been affected by both protagonists of the war.[35]

Table 2.1 shows the official picture of land appropriation from 1945 to the completion of Land Reform. It illustrates the different character of the periods and demonstrates one reason for proceeding with major agrarian reform in 1953. By this time, although the majority of land from sources outside of the resistance coalition had been reallocated--on either a temporary or permanent basis--the demands of the poor peasantry had by no means been satisfied. The shortage of cultivated land per capita presented a more fundamental obstacle to the complete satisfaction of the demand for land.

During the first period, 1945-49, the main source of land was communal, although this probably involved a revival of the practice of periodically reallocating communal holdings and would have been unlikely to have seriously threatened the position of the dominant rural classes. Almost two-thirds of French-occupied land was redistributed during this early phase while providing only 10% of the total for the period. Part of this may already have been regained prior to 1945, and Table 2.1 clearly shows the very limited potential offered by this source. Practice during this period thus followed the policy laid down by the 8th Plenum of the 1st ICP CC in May 1941.[36]

The second period, 1949-53, shows a substantial increase in the amount of land appropriated both on account of the greater effectiveness of the policies and their implementation, and of the increasing area under DRV administration. By 1953 over half of all land appropriated during the entire agrarian reform period had been obtained. The main source remained the communal lands of which almost 80% had been redistributed by 1953. Of the land held by 'traitors' or that uncultivated by its owners, 85% and 75% respectively had been reallocated by 1953. These had, however, provided less than 14% of all land appropriated since 1945.

This may be interpreted in a number of related

Table 2.1: Land Appropriation in the DRV, 1945-57

Original Owner-ship	1945-49			1949-53[1]			1953[1] to LR			During Land Reform			Total		
	Area (Ha)	% (a)	% (b)	Area (Ha)	% (a)	% (b)	Area (Ha)	% (a)	% (b)	Area (Ha)	% (a)	% (b)	Area (Ha)	% (a)	% (b)
French	18,400	61	10	8,400	28	3	2,760	9	2	440	2	---	30,000	100	4
Landlords															
- total	39,600	10	22	117,000	31	39	57,690	15	52	166,010	44	75	380,300	100	47
- traitors	12,850	34	7	18,980	51	6	4,370	12	4	1,370	4	1	37,570	100	5
- unused land	10,570	23	6	23,030	50	8	9,840	21	9	2,790	6	1	46,230	100	6
- donated							2,980[2]	83	3	590	17	---	3,570	100	---
- remainder[3] (scattered, sold, etc.)	16,180	6	9	74,980	26	25	40,500	14	36	161,260	55	73	292,920	100	36
Church land	-	-	-	3,200	13	1	8,440	35	8	12,360	52	6	24,000	100	3
Communal land	119,000	32	67	170,300	45	57	43,050	11	38	43,350	12	20	375,700	100	46
TOTAL	177,000	22	100	298,900	37	100	111,940	14	100	222,160	27	101	810,000	100	100

Notes: (a) Land appropriated from this source as % of the total land appropriated from this source during the whole period, 1945-57.

(b) Land appropriated from this source as % of the total land appropriated from all sources during the sub-period in question.

1 April 1953 in liberated areas, and July 1954 in other zones.
2 This is the total from 1945 to 1953.
3 This category includes the majority of the land appropriated during the Land Reform and Rent Reduction campaigns of 1953-56.

Sources: Table derived from data in Hanoi (1960c), opposite p.66 Table A, and (1978), p.95, and Tran Phuong (1968), opposite p.70.

ways. First, the resistance genuinely included a broad alliance of all classes with very few 'landlords' acting as 'traitors'--the land classed 'unused' would have belonged to those residing in the cities or in other French-controlled areas, and whose attitude to the DRV was less easy to categorise. Secondly, the rural classes, who were at best ambivalent towards the DRV, were not challenged either economically or politically by the DRV and Viet Minh during this period for fear of the resultant disturbances in rural areas and the potential power of the groups involved, or because the Viet Minh lacked confidence in its own strength. In addition the Viet Minh itself was not sure of its future policy towards the rural areas and to 'landlords' and 'rich' peasants in particular. This is not intended to doubt the earlier resolutions of the VWP CC or Politburo, but rather the degree to which they were held with conviction by the VWP and Viet Minh membership at large, due partly to those members attracted during the course of the struggles against both Japan and France.[37] An explanation along these lines would agree with the lengthy period of reform and re-education within the VWP prior to the implementation of the 1953-56 agrarian reforms. However, by 1953, 41% of all land ultimately appropriated from 'landlords' had been obtained. This may have included the larger landowners who had tended to support the French,[38] leaving only small 'landlords' to be appropriated.

The land categorised as Church and that which was 'donated' to the peasantry for distribution were of minimal significance. This left as the major sources of land that which belonged to 'landlords' and perhaps certain 'rich' peasants. The remainder category in Table 2.1 is not completely satisfactory since it includes land scattered--illegally, according to subsequent DRV decrees--and sold in order to preempt denunciation and dispossession during the reforms. In addition it includes the bulk of the land confiscated and requisitioned during the Rent Reduction and Land Reform campaigns by the village committees. The increase in the amount in this category in the second period illustrates the tendency for landlords to disperse their land once it became clear what the future held.[39]

By 1953, the policy of differentiating the 'landlords' had reached the point at which, having expropriated the 'traitorous' and 'reactionary' elements, the main element of the 'landlord' class still existed to be confronted and destroyed both economically and politically. Thus, although 59% of the total land redistributed during the whole period had been redistributed by 1953, this does not diminish the political significance of the major reforms. In comparison with that which remained in 'landlord' hands, this land had been relatively easy to obtain; the rest required a mass mobilisation of the peasantry. Furthermore, that proportion redistributed by 1953 had not changed the agrarian class relationships and had by no means satisfied the rising demands and aspirations of the peasant masses.[40]

Tables 2.2 and 2.3 illustrate the changes in land ownership by class over the course of the period.[41]

Table 2.2: Class Proportions and Land Ownership in the DRV, 1945-53 (percentages)

Classification	1945 in class	1945 land owned	1953 in class during LR	1953 land owned pre-LR[1]
French	---	2.4	---	---
Church	---	1.5	---	1.3
Landlord	4.0	24.5	2.3	18.0
Rich Peasant	2.9	7.1	1.6	4.7
Middle Peasant	34.8	29.0	36.5	39.0
Poor Peasant	37.0	10.0	42.9	25.4
Landless Peasant	17.1	1.1	13.1	6.3
Other Labouring Classes	4.2	0.8	3.6	1.0
Communal Land	---	23.6	---	4.3
	100.0	100.0	100.0	100.0

[1]See note 1, Table 2.1.

Source: Hanoi (1978), p.94. Data collected pre-Rectification.

Table 2.2 shows a reduction in the proportion of 'landlords' and 'rich' peasants as a consequence of the redistribution and the departure of many to French-controlled zones. A less than proportionate fall in holdings of these groups is also indicated. Redistribution of these and the communal lands had the effect of raising some landless peasants into the 'poor' peasant category and of increasing their per capita holdings. The initial level of these holdings, however, was extremely low. A plausible interpretation is that the redistribution raised a number of 'poor' peasants to 'middle' peasants in terms of landholdings, but that classifications remained unchanged since they were based on the amount of owned rather than used land. In principle, the classification used during Land Reform was based on the method of exploitation of this land. Social classification would involve consideration of land, labour credit, and commodity markets.

The figures indicate a similar occurrence among the 'middle' peasant category which gained 10% of the land but only 1.7% of the families during this period, suggesting that some of those classed as 'middle' peasants may have had the real economic status of 'rich' peasants by 1953 by virtue of the fact that their landholding included dispersed 'landlord' and communal land. This would certainly agree with the pressures for infringement of the interests of 'middle' peasants during the Land Reform--especially its later 'leftist' phase--with 'poor' peasants arguing that some 'middle' peasants had use of considerably more than a 'fair' share of the available land. The origins and consequences of these pressures are discussed more fully in Chapter III.

Table 2.3 gives a more detailed breakdown of the sources of land owned and used by the different classes, in addition to figures for per family landholdings in 1945 and 1953. Table 2.2 treats communal land as being distributed to new owners rather than allocated for use as in Table 2.3. The latter treatment is more in accord with both traditional Vietnamese practice and DRV regulations, which up to 1953 had always stressed the provisional allocation of land for a number of

Table 2.3: <u>Class Proportions, Land Ownership and Land Use per Family in the DRV, 1945-53</u>

	1945			1953					land per capita (3)		
Classification	% in class	% of land owned	ha. per family	% in class	% of land owned (1)	ha. owned per family	% of land used (2)	ha. used per family	1945	1953 owned	1953 used
French	---	0.8	---	---	1.0	---	0.3	---	---	---	---
Church	---	1.6	---	---	1.6	---	0.6	---	---	---	---
Common land	---	24.6	---	---	22.5	---	1.4	---	---	---	---
Landlord	3.5	22.5	4.958	1.8	17.3	5.786	10.0	3.352	9741	11368	6585
Rich peasant	2.2	7.1	2.437	1.2	4.4	2.259	4.5	2.288	4423	4099	4153
Middle peasant	32.2	30.4	0.720	33.6	34.6	0.637	45.8	0.844	1549	1371	1816
Poor peasant	35.4	10.8	0.234	43.5	15.7	0.224	29.7	0.424	589	563	1068
Landless peasant	20.6	1.2	0.046	14.4	2.1	0.092	6.3	0.272	145	287	851
Other rural workers	6.1	0.7	0.094	5.3	0.8	0.094	1.2	0.141	244	242	365
Total/Average	100	99.7	0.762	99.8	100	0.620	99.8	0.620	1860	1513	1513

<u>Notes</u>: (1) Counting land which 'landlords' and 'rich' peasants had dispersed as still belonging to these classes.
(2) Counting land which 'landlords' and 'rich' peasants had dispersed as being used by its recipients.
(3) Square metres. Using average family size figures derived from data in Moise (1977), p.419.

<u>Source</u>: Hanoi (1960c) opposite p.66, Table B. Sample of 93 villages and 31 hamlets in 16 provinces, data collected post-Rectification.

reasons: so as not to distort classifications; to retain the possibility of switching land again at a later date; and as a way of dealing with land temporarily abandoned by its owners and cultivators.[42]

With regard to the allocation of communal land, Table 2.3 is in accord with the points noted above, namely, that it appeared to be benefitting 'poor' and 'middle' peasants equally, although the 'poor' peasants might have been expected to receive more given their lower initial per capita holdings. Even 'landlords' seemed to be receiving an appropriate per capita share in the common land --with 1.8% of the families, they were using 2.2% of the common land in 1953.[43] With respect to 'middle' peasants in particular, the table appears to validate the point that a number of them may have attained 'rich' peasant status prior to the major agrarian reforms. The data on illegal scattering of lands by 'landlords' in 1953 shows that this practice involved some 8% of the land and 45% of 'landlord' holdings.[44] Land tenanted or dispersed by 'rich' peasants involved only 8% of their holdings, a negligible amount of the total. Such practices were condemned since they allowed 'landlords' to disguise their class status and allegedly involved clandestine payments of various forms, especially in the early years of the resistance.[45] Towards the end of the period (1951 and 1952) when it became clear that major reforms were to be implemented, this practice of scattering land increased,[46] causing subsequent analytical problems for the Land Reform cadres.

As shown in Table 2.3, average land available per family had fallen by 19% between 1945 and 1953 --the consequence of a 23% rise in the number of families and almost no change in the total cultivated area. Since the increase in population was significantly less than the increase in the number of families, then family fragmentation must have increased and the availability of land per capita fallen by less than 19%. This can be considered in relation to the overall data for the period 1939-54 which show production fluctuating around an average of 2.3-2.4 million tons of paddy, with an average yield of 1,300kg per hectare, and a per capita production of between 180 and 200 kg.[47]

Given the estimated per capita paddy requirement of 240kg[48] per year, these figures show a deficit of some 50kg. The production figures used here may be consistently under-estimating production since peasants sought to avoid taxation and other exactions by under-reporting.[49]

Further, these figures suggest a minimum per capita requirement in terms of land, of 0.18 hectares assuming only one crop per year and no taxation obligations. On the basis of the 1951 agricultural tax, in order to retain 240kg of paddy a family member would require 275kg from owned land, or slightly less from rented-in land. The area required to produce this income would hence become 0.21 hectares, assuming only one crop.

The figures in the final column of Table 2.3 show that in 1953 mean 'middle' peasant land usage was barely sufficient to achieve this basic production requirement. This suggests that any marketed surplus would be derived only from those classed as 'rich' peasants, which illustrates the economic significance of 'rich' peasants and 'landlords' for the resistance struggle.

Table 2.3 shows that 'rich' peasants as a group lost a greater proportion of land between 1945 and 1953 in terms of ownership, than 'landlords': 38% as opposed to 23%. These figures suggest either that 'rich' peasants were unable to withstand political pressures or, more likely, showed a greater susceptibility to economic pressures over this period. In 1953 'landlords' are shown to have lost almost 50% of their land, at least temporarily. The fall in the number of families in these categories, from 5.6% to 2.5% occurring largely between 1945 and 1953,[50] meant that the ratio of land used in 1953 to that owned in 1945 shows less of a fall--32% for 'landlords' and 6% for 'rich' peasants. The individual families who remained in these classes had thus largely maintained their positions. The weakening of the 'landlord' class had consisted of a large number of them being reduced to the ranks of the peasantry rather than a proportionate reduction in the land owned by all those considered as 'landlords' in 1945.

The data in Table 2.3 show higher per capita 'rich' peasant holdings, in 1953 in particular, than in Table 2.4 below. The difference may be due to problems of classification: either some 'landlords' are included in the 'rich' peasant category or some poorer 'rich' peasants are included in the 'middle' peasant category--both of which would raise the 'rich' peasant average. They again illustrate the limited accuracy of the data. In addition, one inconsistency of the Land Reform data is reflected, namely the fall in 'rich' peasant class and per capita landownership between 1945 and 1957 without any appropriation from that class appearing in the summary data presented in Table 2.1.

Table 2.4: 'Rich' Peasant Landownership and Use in the DRV, 1945 and 1953
(square metres)

1945		1953	
owned	used	owned	used
3975[1]	4679[1]	3345[1]	----
4200[2]	5040[2]	----	----
----	----	2141[3]	----
----	----	2116[4]	----
4423[5]	----	4099[5]	4153[5]

Sources:

1) Land Reform Committee 15/8/58--sample of 112 villages and 31 hamlets collected post-Rectification, Tran Phuong (1968), p.14 and Vo Nhan Tri (1967), p.201.

2) Hanoi (1978), p.96, post-Rectification data.

3) Nguyen Duy Trinh to National Assembly 4/1/57--pre-Rectification sample--Shabad (1958), p.46.

4) Nguyen Xuan Lai (1957), p.36, pre-Rectification data.

5) Table 2.3.

As a class, 'rich' peasants appear to have 'lost' some 78,000 hectares between 1945 and 1957. Table 2.3 shows that part of this occurred during 1945 and 1953 as a result of changes in the socio-economic position of many former 'rich' peasant families. The inconsistency occurs then during the period 1953-57, when the classification of 'rich' peasants remained unchanged--1.6% of the population and 1.2% of families--yet their land-ownership fell from 4.4% to 1.8%.[51] The per capita figures shown in Table 2.4 imply a steady weakening of the economic position of 'rich' peasant families from 1945 to 1957, particularly up to 1953. However, this does not correspond with 'rich' peasant class landownership in 1953 of around 4.5%. The possibility that the difference between figures in Tables 2.3 and 2.4 arises from the average family size figures used for conversion seems remote since these show 'rich' peasant families to be larger than all other categories.

In reality, it would appear that a loss in land by the 'rich' peasants was disguised in the official summary data of appropriation (Table 2.1) for political reasons, namely, the need to maintain the concept of a broad class alliance. In addition, the size of the 'rich' peasant strata was consciously underestimated to maintain the notion of an extremely large 'poor' and 'middle' peasant grouping, embracing well over 90% of the peasantry. The continuing problems of defining 'rich' peasants and their landholdings are further discussed in the following chapters.

The conversion of figures in Table 2.3 to a per capita basis overcomes the distortions resulting from the correlation between family size and landownership. These figures are shown in Table 2.5 on a relative basis in an attempt to indicate the degree of relative inequality--on an absolute basis even 'landlord' holdings were very small.

These figures show that substantial changes had occurred between 1945 and 1953, but that inequality remained in terms of access to the major productive resource. Clearly a substantial proportion of the peasantry had seen a temporary improvement in their situation--if the conse-

Table 2.5: Relative Per Capita Landownership in the DRV, 1945 and 1953

	1945	1953 owned	1953 used
Landlords	1645	2019	617
Rich Peasants	751	728	389
Middle Peasants	263	244	170
Poor Peasants	100	100	100
Landless Peasants	25	51	79
Other Rural Workers	41	43	34

Source: See Table 2.3. The holdings of 'poor' peasants have been set at 100.

quences of the war and its attendant instabilities are ignored--and some perhaps saw a potentially more permanent improvement, raising them to 'rich' peasant status. The 'poor' peasant holding, however, was insufficient for producing basic consumption requirements and, for the mass of the rural population, the situation remained unstable even in the best years. Pressure for further reforms could be expected from the peasantry, particularly since their hopes had been raised and they had borne a large part of the burden during the resistance.

'Landlord' holdings, maintained in terms of ownership, had shrunk in terms of usage, and their numbers had dwindled. However, 15% of the land was still owned by 3% of the population, while a further 8% had been temporarily dispersed, and some 22% remained at the disposal of the village-commune authorities--at least in theory. Thus, not only did the 'landlord' class retain an economic and political base in 1953, but the possibility still existed that, given certain circumstances, they could restore their former position.

Summary

Until 1953 the resistance war against France was considered of paramount importance and required the maintenance of a broad united front, although in 1951 the VWP leadership had determined de facto to abandon this policy and proceed to major agrarian reform. Within the nationalist struggle a certain amount of land was redistributed, rents and interest rates were reduced, and a new, severely regressive taxation system was introduced which improved peasant conditions somewhat.

To some extent these measures had assisted the mobilisation of the population against the French but had not involved a decisive attack on the landlord class which retained a powerful economic and political position in rural society. The measures had 'bought' peasant support for the Viet Minh rather than for any other form of government.

Major agrarian reform had not occurred for several reasons. The VWP leadership believed that it lacked sufficient control over the rural areas, and perhaps over many VWP and Viet Minh cadres; there was an absence of sufficiently trained, educated cadres for the campaigns involved; the VWP had to remove other opponents before confronting the 'landlord' class as a whole; and, there was a short-term need to maintain the support, or neutrality, of the majority of 'landlords'.

When these conditions began to change in 1951--decisively so by 1953--mobilisation of the 'poor' and 'middle' peasants against the 'landlords' began. The pace, to a certain extent, was forced by the rising frustrations of sections of the peasantry who had seen only partial satisfaction of their aspirations. These peasants wanted land. They had supported the VWP and the Viet Minh and cooperated in the war because these groups had promised land. To maintain their support the VWP had to show that the promise would be fulfilled.

The Viet Minh had emerged in 1945 after the Japanese coup as the only force capable of filling what was effectively a vacuum in the rural areas. It then had to consolidate its position in terms of its organisational and military strength, as well as in the embodiment of the 'nationalist' struggle. Within this, the ICP/VWP also had to consolidate its position. In doing so, it endeavoured to inculcate the belief that the primary source of social, economic, and political inequality was the old political order which hence had to be transformed. It was in this sense that the post-1945 education and literacy campaigns were of such vital importance.

The period up to 1951 was thus important in spreading the strength and influence which allowed the VWP to 'capture' the nationalist movement. The years to 1953 then saw a further consolidation of this central position: the purification of the movement itself, the training of sufficient cadres to allow 'orderly' agrarian reform campaigns,[52] and a demonstration that, despite the expropriation of land from the French, church, 'traitors,' and the redistribution of communal and abandoned land, the elemental rural power structure remained intact.

Furthermore, this power structure could only be transformed by a major mobilisation, not by administrative or legal measures which so far had had limited effect. The fact that agrarian reforms and land redistribution <u>alone</u> could not produce landholdings sufficient to guarantee minimum production and consumption requirements, subsequently became the focus of attention and required serious consideration. However, by 1953 the clear belief of the VWP (and of the peasantry also, it would appear) was that no other alternative policy, such as consolidation of existing dispersions and guarantees of usufruct rights, could even begin to confront the economic and political problems of the agrarian economy. Such measures would satisfy neither the demands of the peasantry for further land redistribution nor the concerns of the richer rural strata whose economic and political power had already been undermined by the land reform and taxation measures. They would be unlikely to tolerate passively further administrative and legal attacks on their position, especially given the emergence of an alternative regime more in keeping with their interests and objectives.

CHAPTER III

The Agrarian Reforms and Their Aftermath 1953-57

The Timing of the Agrarian Reforms

The agrarian reform policies introduced in 1953 envisaged a two-stage process: first, a campaign of Rent Reduction during which existing legislation regarding rents and interest rates would be rigidly enforced, population classification carried out, and 'traitorous and cruel landlords' isolated; and second, Land Reform itself during which all 'landlord' holdings would be confiscated or requisitioned, and redistributed to the 'poor' and landless peasants. All temporary redistributions of the previous eight years would be checked and ratified during the latter campaign.

Chapter II described the major reasons why 1953 was the year in which these campaigns were launched. First, the momentum for Land Reform had built up, fueled by the rising aspirations of the peasantry--the result of an increasing awareness derived from the Viet Minh education and literacy campaigns of the previous years. Second, all sources of land other than the bulk of landlord holdings had already been redistributed by 1953 without producing sufficient improvement in peasant conditions to satisfy their aspirations and demands. Third, existing legislation had not been fully, or even partially, implemented in all areas due to the retention of considerable power and prestige by some 'landlords' and their supporters. Fourth, the strength of the VWP had increased following the purges of the previous year. Fifth, the military situation had improved, the strength of the army had increased, and base areas had been consolidated, although more support was required for the final steps to victory over the French. Finally, the agricultural tax introduced in 1951, by reducing the incomes of the 'rich', had increased their actual or latent hostility towards the DRV while the nascent Bao Dai government presented a more attractive alternative for them. These campaigns were also seen as a means for destroying pre-capitalist production relations and establishing the preconditions for the transition to socialism, as well as for eliminating remaining opponents to the VWP monopolisation of power.

In purely productive terms, the earlier need to maintain support of the richer rural groups in order to obtain sufficient foodstuffs for the army and administration was being replaced by a belief in the greater productive potential of a small-scale peasant agriculture over what was considered a semi-feudal system. This belief was proven valid by the performance of agriculture in subsequent years, although agricultural production would inevitably have increased following the end of hostilities and attendant rural disruption. The growth possibilities would depend on whether increased production was used to raise agricultural consumption, agricultural accumulation, or industrial accumulation, as well as on the incentives for continued expansion of agricultural production. Longer-term growth would require increased inputs from the industrial sector.

This combination of factors required a campaign in which the economic and political positions of the richer rural strata would be destroyed, while still maintaining a united front of the other sections of the rural population and thereby causing minimum disruption to production. However, although this provided sufficient rationale for the agrarian reform campaigns, it was also potentially contradictory. The limited absolute amount of land still held by 'landlords' was insufficient to satisfy 'poor' peasant demands and produce viable holdings for all the rural population. Furthermore, in previous years, the policy of a united front had resulted in the Viet Minh, VWP, and other organisations including the rural rich among their memberships--often in dominant local positions.

A policy aimed at destroying these groups would thus place enormous strains on rural organisations. This became particularly apparent given the methods used during Land Reform: the reliance on relatively poorly trained cadres and the application of a somewhat simplistic class policy to a complex rural structure.

At the 4th Plenum of the 2nd VWP CC, (25-30 January 1953[1]) the arguments for a two-stage campaign were made clear: to differentiate the 'landlord' class and to initially overthrow only the most reactionary elements within it; to raise peasant consciousness further and prepare the mass of the peasantry for the major campaign against the whole 'landlord' class; and to help train cadres and accumulate experience for the subsequent, more difficult campaign. However, it was recognised that in places the two campaigns could be combined--an occurrence that became quite widespread: 1,875 villages completed Rent Reduction and 3,314 Land Reform.[2]

The VWP CC meeting was followed by a joint meeting of the National Assembly Standing Committee and the National Committee of the Lien Viet (the united front organisation which had developed from the Viet Minh) held from 25 February to 1 March 1953. The major speech on agrarian policy at this meeting was made by the VWP Secretary General, Truong Chinh, who called for both the full implementation of existing land policy and the government's formulation of further legislation.[3] His theme was taken up by other speakers, including Ho Chi Minh and Pham Van Dong, who referred to the failings of cadres and the 'infiltration' of local committees and peasant associations by 'landlords' and their agents as factors contributing to the limited implementation. Pham Van Dong made it clear that the aims of the forthcoming campaigns were both to satisfy the economic needs of the peasant masses but,

> also and essentially, to succeed in mobilising the peasant masses to ensure their supremacy on the political plane where they have to overcome domination of the landlords and other powerful persons.[4]

These speeches clarify the decisive shift in policy (announced in 1951) whereby the united front and membership of mass organisations were no longer to include 'landlords'. These reports also repeat the argument, used by Truong Chinh and others for some time, that mass mobilisation for

Rent Reduction and Land Reform was necessary to both consolidate existing support and gain new support for the final defeat of the French.

One additional reason for the timing of these measures in 1953 may be related to the completion of the Chinese Land Reform in late 1952. Ho Chi Minh, in his speech to the National Assembly meeting which passed the 1953 Agrarian Reform Law, commented that,

> the experiences of other countries have taught us that a successful Land Reform will help us to overcome many difficulties and solve many problems.[5]

In the same week that this speech was published it was reported that articles written by Mao Tse Tung between 1926 and 1945 had been published and should be studied by cadres.[6] These would presumably have included 'Analysis of the Classes in Chinese Society' (March 1926) and 'Report on the Investigation of the Peasant Movement in Hunan' (March 1927).[7]

A further reason for the timing of the agrarian reforms was the introduction of a land policy by the French-supported Bao Dai government in 1953.[8] This policy was also nominally based on the principle that land should belong to those who cultivated it, and it fixed maximum landholdings at 100 hectares in southern Vietnam, 45 hectares in central Vietnam, and 36 hectares in northern Vietnam. The limited significance of this policy is illustrated by the fact that equal land distribution throughout Vietnam would have given each peasant family one hectare, and in Tonkin under two-thirds of a hectare.[9] The Bao Dai policy fixed the maximum land rent at 15% of total yield, provided for cheap loans by agricultural cooperatives--whose development it encouraged--and land banks, and created mechanisms for distributing abandoned and reclaimed land. Although this policy was never implemented, it conceivably could have provided a counter-attraction to existing Viet Minh policy in disputed territories.

Early Phases of the Rent Reduction Campaigns

The legislation to launch the Rent Reduction campaign consisted of a Class Classification Decree of 2 March 1953, the Decree of Agrarian Policy on 12 April, and an implementing Decree of 20 April.[10] These decrees aimed to consolidate all existing legislation relating to rent and interest reduction while ensuring its enforcement and, in so doing, to fix the class classification in the rural areas. In addition, the land of the 'reactionary and cruel landlords' was to be expropriated.

The numerous existing decrees relating to land and temporary distribution had been enacted at different dates and implemented to varying degrees in different areas. Moreover, since 1940--and particularly since 1945--the degree of instability had increased greatly and land had been dispersed and temporarily redistributed. The April legislation thus provided a basis from which to proceed: no forms of supplementary rent could be collected, many debts were annulled, and no further dispersions of land could occur. Indeed, many dispersions which had occurred since the earlier 1949 legislation were declared illegal and were thus liable for redistribution.

The first wave of Rent Reduction was launched in twenty-two villages between April and August 1953,[11] although trials may have occurred in certain villages from late 1952.[12] The first wave of Rent Reduction was discussed at a VWP Politburo meeting in August 1953[13] and revealed problems of implementation. These problems which essentially related to the class line and the scope of attack envisaged, re-emerged throughout the Rent Reduction and Land Reform campaigns.

The Politburo resolution stressed the need to maintain a broad united front and confine the attack to only the 'most reactionary landlords' who had acted for the French against the policies of the DRV and Viet Minh, and to those who were guilty of serious crimes such as murder and rape. By dealing with the 'ringleaders,' even within these categories, between one and five 'landlords' were to be 'struggled against' in each village depending on its size, rather than ten to fifteen as was the case in the early wave. The Politburo stressed that the class line laid down by the VWP must be carefully adhered to by the implementing cadres and not be violated with respect to 'middle' and 'rich' peasants.

Although based on the 'poor and landless' peasants, this line involved uniting closely with the 'middle' peasants and allying with the 'rich' peasants in order to overthrow the 'reactionary Vietnamese traitor landlords' and 'cruel notables.' The policy towards the 'rich' peasants stressed that they were free to hire labour and could take part in the campaign against the 'landlords.'

The significance of the 'rich' peasants and even 'patriotic landlords' was clearly both economic and political. Their contribution to agricultural production was considerable and its maintenance was essential to ensure reasonable stability of production. A political campaign against them would widen the struggle to a dangerous degree, particularly given the classification difficulties. Thus the VWP leadership wished to ensure at an early stage that the policy was not exceeded--though its ability to do so was clearly limited. From the point of view of the Land Reform cadres, relying on evidence of oppressed peasants and landless agricultural labourers, the distinctions between different categories of 'landlords' and 'rich' peasants were frequently extremely difficult to discern. This was especially true with respect to the distinctions between peasants who hired labour and had a much more direct relationship with their workers--and hence appeared more oppressive and 'cruel'--than that which a large landlord might have with his tenants. This issue is discussed later in this chapter when classification and its problems are examined more fully. Here it is significant to note the very early emergence of these difficulties.

The available data relating to the twenty-two villages which first experienced Rent Reduction indicate a very wide attack. Figures for all 1,875 villages undergoing Rent Reduction show an average of 2.1 'ringleader reactionary landlords'

--8.8% of those classified as 'landlords', and 3.8% 'cruel notables'--16.2% of the total classified as 'landlords.' Thus, in each village during Rent Reduction, some six 'landlords' (representing 25% of all 'landlords') were 'struggled against,' whereas in the twenty-two first-wave villages, between ten and fifteen were opposed, representing approximately 50% of all 'landlords.'[14] By way of comparison, the post-Rectification classification considered only 13% of 'landlords' to have been in these categories.[15]

Figures showing the class composition of the local VWP committees in these villages also reveal a dramatic change. Prior to Rent Reduction the composition had been 'landlords',13.5%; 'rich' peasants, 15%; 'middle' peasants, 61.4%; 'poor' peasants, 3.7%; and 'petty bourgeois', 3.5%. Following the campaign and the expulsion of 'landlords' and 'rich' peasants, the proportions then became 'middle' peasants, 44%; 'petty bourgeois', 3%; and 'poor' peasants and 'landless labourers', 53%.[16] These figures suggest that approximately half of the membership of these VWP executive committees changed after Rent Reduction, clearly indicating the extent to which the existing local leadership was mistrusted by the campaign cadres and the peasants they mobilised.

It is possible that the membership did not change to this extent, but that 'landlord' and 'rich' peasant members had taken measures to change their class classification. The use of cadres from outside may have been an attempt to overcome this or may have made it easier. It is unlikely that 'landlords' and 'rich' peasants successfully disguised their status to any great extent since the campaigns tended to overestimate numbers in these classes.

These figures also illustrate the degree of 'landlord' and 'rich' peasant involvement in the VWP itself during the resistance, suggesting that their role in the Viet Minh and other organisations may have been even greater. The figures illustrate both the success of the united front policy and the problem that that very success posed for the anti-'landlord' campaign.

The implementation of Rent Reduction[17] involved 'work teams' of specially trained cadres entering individual villages. They were to publicise the policy and discuss it with the local VWP cell and village committee before 'striking roots' among 'poor' and landless peasants with whom they lived, ate, and worked, and who did not have to be VWP members or village cadres. These peasants were encouraged to discuss their living conditions and their sufferings at the hands of 'landlords.' It was explained that these conditions resulted from the system of landlordism rather than from the oppressive practices of individuals, and hence required the overthrow of the system. These 'poor' peasants in turn discussed with other 'poor' and then 'middle' peasants until sufficient support existed for a mass public meeting at which the 'ringleader reactionary and cruel landlords' were denounced publicly.

After the first wave of Rent Reduction and the Politburo meeting in August, those denounced were judged and sentenced in Special People's Courts rather than at the mass meeting itself. Denunciation asserted the supremacy of the new class line ('poor' and 'landless' peasants united with 'middle' peasants and in alliance with 'rich' peasants). After this came the measures covering rent and interest reduction, collection of unpaid wages, and (temporary) distribution of appropriated land and other property. The final stage involved the reorganisation and consolidation of the VWP and other administrative organisations. Land Reform followed a pattern similar to Rent Reduction except that the final stage involved the permanent distribution of the majority of existing landlord holdings.

The repayment of excess rent and interest applied both to 'landlords' and to 'rich' peasants whereas, after the first wave of Rent Reduction, payment of previously unpaid wages was confined to those denounced during the Rent Reduction campaign. Distribution and repayment gave priority to those who had paid the excess rent or interest or received inadequate wages; to those to whom land had been temporarily redistributed or dispersed; and to those most in need. These were not always easy to reconcile.

The second wave of Rent Reduction was carried out from August to October 1953 in 163 villages in ten provinces in the light of the Politburo resolution.[18] In November the final preparations for Land Reform were made, and thereafter the two campaigns proceeded alongside each other.

The Land Reform Campaign

The promulgation of the Land Reform Law on 19 December 1953[19] (effective from 1 January 1954) followed a series of major meetings during the course of November 1953.[20] These began with the 5th Plenum of the 2nd VWP CC at the beginning of November which laid down the objectives of the campaign. These were wider than the objectives of the Rent Reduction campaign in a number of respects. The entire landlord class was to be overthrown, thorough-going reconstruction of all local and VWP organisations was to take place, as was the permanent distribution of ownership rights in land in most areas (the only exceptions were certain suburban districts).

The VWP CC Plenum was followed by a National Conference of the VWP from 14 to 23 November at which the leadership policy was presented by Truong Chinh.[21] In terms of the two-stage revolution, he argued that with the now almost certain success of the 'anti-imperialist' struggle, the 'anti-feudal' struggle could begin, although he also repeated his earlier contention that the latter had already been postponed for too long. This meeting was followed by a conference of the National Committee of the Lien Viet (26-28 November) which approved the VWP programme, and by the 3rd Session of the National Assembly from 1-4 December which unanimously approved the law subsequently promulgated on 19 December. The order of these meetings confirms the VWP monopoly of power and the minimal role assigned to the National Assembly, which had not met since November 1946. Its meeting was attended by only 171 of its original 444 members,[22] a large proportion of whom presumably did not dare to attend.

The December decree did not differ substantially from those of March and April[23] and with

Table 3.1: Waves of Rent Reduction and Land Reform

Rent Reduction Wave	Land Reform Wave	Dates	Villages RR	Villages LR
1		April - May 1953	22	---
2		Aug. - Oct. 1953	163	---
3	Experimental	25/12/53-30/03/54	...	6
4			572	---
5	1	25/05/54-20/09/54	...	53
6	2	23/10/54-15/01/55	255	210
7	3	18/02/55-20/06/55	344	466
8	4	27/06/55-31/12/55	519	859
	5	25/12/55-30/07/56	---	1720
			1875	3314[1]

Note: 1. Subsequently divided into 3,653 smaller villages.

Sources: Land Reform data from Tran Phuong (1968), pp.107-8, also Moise (1977), pp.299-301. Rent Reduction data from various contemporary reports cross-checked for consistency, total from Tran Phuong (1968), p.111.

these formed the body of legislation for Land Reform. Subsequent amendments were made in March 1955 to the Population Classification regulations, and in May and June 1955 to the Land Reform law.[24] These amendments resulted partly from the implications of the Geneva agreements which had been discussed at a VWP Politburo meeting in September 1954. They also reflected further concern over the wide-ranging nature of the attack in many areas and the need to maintain an orderly approach with minimal disruption.

The experimental wave of Land Reform took place between December 1953 and March 1954 in six villages in Thai Nguyen province and was followed by the first wave in 53 villages which was completed in September 1954. By this date, after the Geneva negotiations and temporary partition of the country, five waves of Rent Reduction had been completed in some 750 villages.

The end of the war and the Geneva treaty meant that the entire northern area was not in the control of the VWP. This seriously weakened the positions of both the 'reactionary landlords' and those with a preference for the Bao Dai regime, while it made Land Reform easier. However, a wide united front was still required to win over the majority for national unification. The basic line of Rent Reduction and Land Reform remained unchanged, although the legislative amendments of 1955 moderated the policy in some respects.

In practice, the changes in the political situation and the modifications in policy seem to have had little impact, with 'deviations' from the official line continuing. During the trial wave of Land Reform these were labelled 'rightist' and 'leftist'. The 'rightist' tendencies consisted of relying on administrative measures more than on class struggle in identifying 'landlords,' and of failing to satisfy 'poor' and landless peasant demands when distributing land and other property. These practices were frequently identified in time and were overcome, whereas those of the 'leftist' variety were more pervasive.

The 'leftist' practices have been discussed with respect to the first Rent Reduction wave. Despite their early identification during both campaigns, they continued to recur for reasons related to the complexity of the class situation in practice, (as opposed to its simplicity in theory), the relative autonomy of the Land Reform cadres, and the belief that class status determines everything.[25]

These 'leftist' deviations consisted of: emphasising only the shortcomings of agrarian reforms during the previous years; underestimating the changes which had occurred in the balance of class forces during this period; and, not recognising the strength of VWP and other rural organisations and overestimating the power of the 'landlord' class within them.[26]

Class Classification[27]

The crucial part of Land Reform was the

categorisation of the rural population. This followed the arrival of the Land Reform cadres in a particular village and preceded the appropriations and redistributions which were based upon it. Classification was based on the hộ--a statistical unit of population translated as 'family unit'. The actual peasant household was often larger than the hộ.

'Landlords' were defined as those who owned 'much' land, did not engage in 'principal' labour, and whose main source of income was from rents and usury. This involved consideration of landownership and use, labour use, and relative sources of income. In terms of ownership, those families who owned, per capita, three times the local per capita average and rented out their land, were classed as landlords. If a family consisted of only one or two members it could own four times the local per capita average. Thus, 'families' who owned relatively little land but also had very little labour power were, theoretically, exempted from the 'landlord' category.

In cases where part of the land was farmed by family labour, the criterion was that the amount rented out exceeded three times that farmed, or, the income from renting was three times that from farming and in excess of 40 quintals of paddy--9.8% of 'landlords' were apparently in this latter category.[28] The labour of secondary wives, adopted children, and other nominal members of the family whose living conditions more closely correspond to those of poor peasants was not recognised as family labour. It was necessary for a genuine member of the family to engage in principal labour (ploughing, harrowing, transplanting, weeding, harvesting, and irrigating) for four months of the year for the family to be recognised as being involved in principal as opposed to secondary labour (care of animals and fruit trees or vegetable-picking). For larger families it was necessary that one-third of its capable labour force be engaged in principal labour. This standard was adjusted for those areas with only one harvest and where the period of principal labour was less than 120 days per year.

Exemptions were also made in the case of

families whose main income derived from industrial or commercial activities, as long as they had been thus engaged for one year prior to 19 December 1953. A subsequent amendment of October 1955 relaxed this to allow for exemption as long as obligations under the law, such as repayment of excess rents, were fulfilled. Priests and other leaders of religious orders which owned land were also not to be classes as 'landlords'.

The classification thus attempted to take into account families with little land and little labour, families with non-agrarian incomes, and families with large amounts of land but also substantial labour power. However, the problematic areas are apparent from this. They included: the absence of complete or acceptable cadastral records on which to base evaluations of landownership; the problem of discerning who actually constituted family members; that of ascertaining whether principal or secondary labour was engaged in; and the question of which base period was to be used for the classification. The latter was important in terms of the instability of the preceding period, the possibility of circumventing legislation and affecting classification by dispersions and more complex letting arrangements, and the likelihood that landowners would shift from one form of 'exploitation' to another depending on the relative demand for land and the price of agricultural labour. These problems are in addition to those related to the peasant family life-cycle, which could result in the elderly renting out most of their land or hiring-in all the labour power required to work it.

The official data in Table 3.2 show an increase in 'exploitation' via rental between 1945 and 1949. This could be because of the drain on able-bodied labour caused by the war, the continuing population pressure on land, the restricted 'landlord' access to communal land which would tend to be cultivated with hired labour, and the elimination of some of the 'landlords'.

The base date for classification was 1949, since exemption was only possible if an alternative source of income had existed, principal labour been engaged in, or renting out had ceased,

Table 3.2: Renting Out Land and Hiring-in Labour, 1945 and 1949

	1945	1949
Percentage of landlords:		
relying solely on renting out	17.8	19.5
relying solely on hiring labour	43.0	30.5
both renting and hiring	39.2	50.0
Percentage of landlords' land:		
rented out	49.7	55.0
worked by hired labour	50.3	45.0

Source: Tran Phuong (1968), p.150. Data from 124 villages in 11 provinces, collected post-Rectification.

for five years prior to the enactment of Land Reform. In newly liberated areas this was subsequently amended to five years prior to July 1954. Thus, all dispersions made since 1949 were nullified except where a family could claim to have been 'ruined' by natural or other factors. Since this was a claim that could plausibly be made by much of the population, it illustrates a further ambiguity.

Table 2.3 showed official data on the extent of the dispersion of land in 1953 and also the changes in 'landlord' proportions between 1945 and 1953, although given the classification system detailed above, 1953 classes are effectively those of 1949. By the time Rent Reduction and Land Reform were carried out--in 1955 and 1956 in the majority of areas--the basis of the classification was some years out-of-date, and dispersions had increased considerably. Moreover, these changes had not occurred uniformly as the data in Table 3.3 show. Certain changes would have resulted from demographic factors while economic conditions --particularly the payment of taxes to both the French and DRV--would have resulted in other 'departures' from the 'landlord' class.

These data illustrate the degree to which the number of 'landlord' families per village was over-estimated in the first Rent Reduction wave, which 'struggled against' ten to fifteen landlords

Table 3.3: 'Landlord' Families, 1945-53

Areas	Sample Villages	Landlord families 1945[1]	Landlord families 1953[1]	Changes in landlord class A	Changes in landlord class B	1953 as % of 1945
Liberated	53	800 (15)	605 (11)	21	219	75.7
Temporarily occupied	35	582 (17)	576 (17)	86	92	98.9
Guerrilla	32	642 (20)	483 (15)	11	170	75.2
Liberated with high proportion of communal land	7	215 (31)	103 (15)	1	113	47.9

Notes: 1. Figures in parentheses are the average number of landlord families per village.
A. Landlords joining the landlord class.
B. Landlords leaving the landlord class.

Source: Tran Phuong (1968), p.157. Data collected post-Rectification. The geographical areas are not specified nor whether the 7 'liberated' villages are included in the 53.

per village, and Rent Reduction as a whole, which classed on average 24 families per village as those of 'landlords'.

The classifications laid down further subdivisions of the 'landlord' category which were mentioned above in the context of Rent Reduction, namely, 'cruel and reactionary Vietnamese traitor landlords,' 'ordinary,' and 'resistance landlords'. These categories were to be accorded different treatment.

In the classification decrees, 'rich' peasants were defined as owning land requiring the employment of labour in addition to their own family labour. The amount of land concerned could vary considerably given the variation in family size. They might also rent out land, but their major source of income was through the use of hired labour. In certain cases they might use hired labour on rented-in land. The distinctions between 'rich' peasants and 'landlords' have been given above. The criteria for distinguishing between 'rich' and 'middle' peasants, who might also hire some labour, were that the former hired the equivalent of two man-years (or 240 man-days) of labour or, rented out land that would require two working peasants to farm, or a combination of these two.

'Middle' peasants were those who relied primarily on their own labour although they might hire-in some labour or rent out some land. 'Poor' peasants were those who had insufficient land to provide minimum consumption requirements for family members and had to rely on renting-in additional land or hiring-out their labour power. 'Landless' peasants had virtually no land or other means of production and were totally dependent on hiring-out their labour power.

A fundamental difficulty of the guidelines was the problem of accounting for changes in family circumstances. Family size was positively correlated with past and current per capita production. Favourable harvests leading to the survival of more children and more labour could result in a land-poor peasant family maintaining per capita consumption levels and renting-in land to provide for an expanded family. A very large, but successful family could thus have a large operational holding but still own 'insufficient' land for its needs.

In contrast a family with an initially large, owned holding, accumulated over time because of favourable conditions, might lose a large number of members through an epidemic or other disaster, and hence find itself with an 'excess' of owned land in relation to its consumption requirements. The guidelines were inevitably ambiguous in attempting to fix the classification at a point in time in a situation that was changing continually as a result of demographic and other factors.

The classification stage was followed by the appropriation of land and its redistribution. This entailed difficulties in addition to those resulting from classification. In theory, different methods of appropriation were to be adopted for different groups.[29] Land owned by non-Vietnamese (mainly French), 'reactionary Vietnamese landlords' and 'traitors and cruel notables' was to be confiscated. Land owned by religious groups and communal land was to be appropriated without compensation, while land owned by ordinary and resistance 'landlords' was to be appropriated with compensation.

Compensation was to be paid in the form of government bonds yielding 1.5% interest annually, repayable in ten years. The value was determined by the 'average' annual production (over an unspecified period) of the land appropriated; for draft animals and implements, the current market price was used. All property and possessions belonging to 'cruel and traitorous landlords' were liable for confiscation, while 'excess' housing and foodstuffs were to be appropriated (with compensation) from 'ordinary landlords' only if insufficient amounts had been obtained from other sources. Given the ambiguity of the classification and the absolute scarcity of land, this provision was open to varying interpretations. In all cases, 'landlord' families were to be left sufficient resources to make a living from their own labour. The post-Land Reform data show that this provision was usually violated.

Land was to be appropriated (with compensation) from 'rich' peasants only in those cases where expropriation of 'landlords' provided insufficient land. 'Middle' peasant-owned-land was to be untouched and augmented where appropriate. The principle of redistributing land was, in essence, simple. All land still held by 'landlords' in addition to that illegally dispersed since 1949--with the exception of plantations and appropriate portions for religious purposes and 'landlords' families--was to be available for distribution. It was to be divided, taking into account varying land productivities, by allocating portions based on yield, not area, and based on the consumption requirements rather than the labour power of individual families. Consumption and labour power varied through the family life-cycle. Therefore, allocations were only appropriate to the particular time of distribution.

The norms for distribution were to be based on the average ownership per family member in the village. Land dispersed illegally during previous years was to be redistributed if it had been dispersed to 'landlords' and 'rich' peasants, but not if it had gone to land-deficient peasants. 'Middle' peasants were liable to lose earlier gains if their current holdings were in excess of the 'middle' peasant average, and/or if there was insufficient land from other sources to raise the average holdings of land-deficient peasants.

The absence of acceptable cadastral records, the extensive sub-division of plots, and the impossibility of assessing averages--of land areas, productivities, and holdings--over time when the number of families and family members was variable, were clearly sources of further ambiguity. In addition, the relevant laws called for a reliance on the peasant masses in estimating these variables. However, the major difficulty concerned temporarily dispersed, or redistributed land. According to the official data reported in Table 2.1, only 44% of 1945 'landlord' holdings remained in their hands by the time of Land Reform while, according to data reported in Table 2.3, 'landlords' had dispersed 45% of their 1953 holdings and were using only 45% of their 1945 holdings. Moreover, much of this land had been distributed to 'middle' peasants and was liable for redistribution during Land Reform.

Table 3.4 uses survey data to carry out a hypothethical application of the guidelines and produces holdings roughly compatible with other post-Land Reform data. Since the guidelines did not provide objective criteria for appropriation or distribution, Table 3.4 presents only one possible interpretation. By using per family rather than per capita data, it is likely to have exaggerated the need to appropriate land used by 'middle' peasants. However, it does not take account of land dispersed between 1945 and 1953. The sample shows 1953 'middle' peasants owning, as

Table 3.4: Exercise in Land Distribution

	1953 families	1953 land owned (ha)	1953 land owned (ha) per family	1955 land owned ± dispersed (ha)	1955 land owned (ha) per family	Round 1 (ha)		Round 2 (ha)	
						Land taken	New distribution	Land taken	New[2] distribution
French	-	420	-	144	-	144	-	-	-
Church	3	687	-	277	-	277	-	-	-
Common	-	9871	-	630	-	630	-	-	-
Landlord	1311	7586	5.7864	4394	3.3516	3581[3]	0.6204	-	0.6204
Rich	854	1929	2.2588	1954	2.2881	-	2.2881	25	2.2588
Middle	23774	15154	0.6374	20073	0.8443	-	0.8443	4919[4]	0.6374
Poor	30706	6863	0.2235	13021	0.4241	-	0.5278	-	0.6204
Landless	10216	936	0.0916	2775	0.2716	-	0.3753	-	0.5259
Others	3736	350	0.0937	582	0.1413	-	0.2450	-	0.3956
Total	70600	43796	0.6204	43850	0.6204	4632	0.6204	4944	0.6204

Notes:

1 Total land appropriated is distributed to landless peasants, poor peasants, and others. 4632 ha ÷ 44,658 = 0.1037 ha per family.

2 Poor peasant deficiency = (0.6024 - 0.5278) ha x 30,706 = 2843 ha. Remainder distributed to landless peasants and other workers (4944 - 2843) ha ÷ (10216 + 3736) = 0.1506 ha per family.

3 Landlords retain average holding of 0.6204 ha, thus, appropriation = 43914 - (1311 x 0.6204 ha) = 3581 ha.

4 Since average middle peasant holding exceeds overall average, then 1953 dispersions are appropriated.

Source: Original data from 91 villages and 31 hamlets, Hanoi (1960c), p.66.

a group, 14% more land than 1945 'middle' peasants.[30] Land Reform also involved the appropriation of more 'rich' peasant land than indicated here--although this was not officially sanctioned or acknowledged. It also involved the appropriation of land from 'landlords' which left their families with substantially less than the average per capita holding.

Table 3.4 implies the appropriation of 25% of land used by 'middle' peasants in 1953, indicating the extent to which the position of this group was threatened regardless of problems of classification and implementation. Moreover, since the above calculations are based upon averages, the threat to those at the top end of the 'middle' peasant category was proportionately greater. This outcome would affect food mobilisation because these peasants were more likely to produce a marketable surplus over and above the consumption requirements of their own family.

The more equal family-owned holdings resulting from Land Reform would be likely to boost production via the once-for-all effect of increased intensity of labour utilisation. The proportion of output marketed might fall, depending on whether the peasantry treated any marketable surplus as residual after satisfying their own consumption requirements, or whether they marketed that amount required to finance the purchase of goods not produced within the family. It would also depend on the extent to which higher income earners had a greater propensity to consume goods produced outside the family, and the degree of egalitarianism in the post-Land Reform wealth and income distribution. The net short-term effect on the marketable surplus would depend on whether the increased output outweighed any tendency for the marketed portion to fall. The longer-term growth implications of the post-Land Reform landholding structure would be influenced by a greater number of factors, including: the possibilities and incentives for accumulation within agriculture; the availability of inputs from outside of the agrarian sector; the terms of inter-sectoral exchange as well as demand and supply constraints on the development of this exchange; and, the extent to which even the short-term output increases exceeded the rate of population growth.

What is clear, therefore, is that the relative scarcity of cultivated land in the Tonkin delta in particular, and in the DRV in general, coupled with the distribution of 'landlord' and communal land in the years since 1945, though possibly also since 1940, meant that the economic position of 'middle' peasants would inevitably come under attack during Land Reform. This occurred independently of any political attack which the implementation of Land Reform involved and of the classification problem--although to some extent all of these factors overlapped. Since the distributions of earlier years had raised the effective status of some 'middle' peasants, they were vulnerable when both additional sources of land were required, and certain percentages of 'landlords' and 'rich' peasants were sought.

According to the data reported in Table 2.3 the percentage of 'landlord' families had fallen from 3.5% of the total in 1945 to 1.8% in 1953, and 'rich' peasant families had fallen from 2.2% to 1.2% over the same period. Data from a different sample are shown in Table 3.5. According to this source, the combined total of 'landlord' and 'rich' peasant families fell from 5.6% of the total in 1945 to 3.5% in 1949, and at most 3.1% in 1953 (assuming no change in the 'landlord' proportion between 1949 and 1953).

Table 3.5: 'Landlord' and 'Rich' Peasant Families, 1945-53 (percentages)

	1945	1949	1953
Landlord - families	3.4	2.0	---
Landlord - population	3.9	2.3	---
Rich peasant - families	2.2	1.5	1.1
Rich peasant - population	2.9	2.2	1.6

Source: Tran Phuong (1968), pp.156 and 160. Data from 112 villages and 31 hamlets collected post-Rectification.

Reasons for this fall have been mentioned earlier and include: the effect of early rent reduction and redistribution measures; the 1951 tax reform; the reduced access of these classes to communal land; and, the impact of the instabilities of these years. In the case of 'rich' peasants, the reduced labour force available as a result of the possibilities of employment on war-related tasks and the partial improvement in the living conditions of 'poor' and landless peasants would have necessitated a greater use of family labour, and hence effective reclassification as 'middle' peasants. The female participation rate in 'rich' peasant families would have risen. It must be remembered that classification was based, in theory, on the method of exploitation rather than on the amount of land owned, although the latter, simpler criterion was frequently used in practice.

If the assumed 1945 classification was used as a basis for Land Reform investigations, then cadres would be expecting to find a much higher proportion of 'landlord' and 'rich' peasant families than actually existed. An attempt to place 5.6% of all families in these categories would necessitate the inclusion of many 'middle' peasant families. Assuming 'middle' peasant families made up the 2.5% deficiency, then this represented almost 10% of all 'middle' peasant families. In practice, these ratios were often treated as a minimum, resulting in an even greater degree of mis-classification of 'middle' peasants. The major official study of Land Reform refers to a belief in landlord proportion of 4-5%.[31] The admission of 'errors' in late 1956 included the recognition that,

> Land Reform cadres wrongly estimated the force of the enemy...as a result a number of poor or landless peasants themselves have been hit, many middle peasants were also affected, rich peasants were regarded almost as landlords.[32]

The violation of classification norms within the 'landlord' class--particularly the exaggeration of the proportion of 'cruel and reactionary landlords' where the maximum stipulation of 25% was treated as a minimum[33]--coupled with the effect of the 'target' percentages, would even have resulted in the possibility in some instances of 'middle' peasants being classed as 'cruel landlords'.

An official classification survey conducted after Rectification revealed 6,220 families in the 'cruel reactionary' category.[34] This represented 13% of all 'landlord' families and less than two per village. The Rent Reduction campaign had claimed to identify six families per village in this category, and the trial wave ten to fifteen.[35] The latter would have included all 'landlord' families as most data suggest average numbers of 12-18 'landlords' per village. In such cases the other 'landlord' categories would presumably have been filled by 'rich' and 'middle' peasants, and even some 'poor' peasants.

Thus, an examination of the guidelines for Land Reform in conjunction with available empirical data from official sources suggests that a substantial infringement of 'middle' peasant interests was inevitable. This was particularly true of that strata within this class which, along with the 'rich' peasants, would have provided the bulk of the surplus production on which the State depended. This infringement derived from the application of an ambiguous class classification and from the shortage of land available. It received an additional impetus from the method of application of Land Reform, in particular its reliance on cadres from outside who came in to mobilise the land-deficient peasants. Untainted by associations with the richer rural groups, these outsiders served as the vanguard cadres of the struggle.

Discussion of the Land Reform has normally concentrated on the methods of implementation and the various phases involved, drawing attention to the radicalisation of practice alongside the moderation of policy as the programme progressed.[36] The argument here emphasises the early 'excesses' of both Rent Reduction and Land Reform and demonstrates their basis in the inherent nature of the Land Reform classifications and regulations.

Further, these potential conflicts actually occurred and were exacerbated because implementation relied on a process of class struggle in conflict with the earlier united front policies. This resulted in, and indeed required, the bypassing of the organisations, including to a large extent the VWP itself, which had developed during the united front resistance period. Not only was it necessary to purge the richer rural strata from these bodies, but also middle and poor peasants who had played a major role in the organisations and had become tainted by that association.

Middle peasants who had almost no economic interest in Land Reform beyond the control it guaranteed them over some of the land acquired in previous years, could not be expected to play a major role in the campaign. However, giving effective power to the poor and landless (and often the most marginal within these groups, such as 'outsiders' and the very young)[37] in the struggle for land also gave them power in the political struggle.

Thus, a far-reaching purge of all existing organisations was undertaken based on the belief that 'landlords' were the 'incarnation of evil',[38] and their influence had been more pervasive in recent years than was actually the case.

> We underestimated our organisations in the villages, and overestimated the control of the landlord class: we did not clearly recognise that during the Resistance war, a rather large proportion of the poor peasants and agricultural labourers had had their ideological and political consciousness raised, and some of the (more) advanced poor peasants and labourers had participated in village organisations. As a result, during the Land Reform we did not emphasise strengthening and relying upon these poor peasants and labourers: we even grew suspicious of them and attacked them...[39]

This quotation illustrates the degree to which the VWP leadership mistrusted the peasantry and yet associated the peasant desire for land with active support for all VWP policies. The VWP had gained peasant cooperation during the resistance and a monopoly of power by accepting the peasantry's demand for land. The organisations in the villages had been primarily peasant, rather than VWP organisations.

There were other reasons for attacking middle peasants, even excluding consideration of the cynical pursuit of power motive. Some middle peasants, like 'rich' peasants, engaged in a more direct form of exploitation than renting out land; namely, hiring labour or using family labour in the guise of secondary wives and adopted children. This was a form of bonded labour, although those involved may have been better off by being taken into the family in this way. Differing interpretations of the classifications relating to principal and secondary labour, as well as motives of revenge, would be sufficient to place a middle peasant in the 'landlord' category, where a failure to differentiate between 'cruel,' 'ordinary,' and 'resistance' could result in severe punishment.

The adoption of a dichotomous view of class,[40] which was inherent in the struggle notion, obscured the subtle distinctions in the actual classification decrees. Thus, those with large landholdings, whether landlord, rich peasant, or large middle peasant families, were grouped together and denounced, regardless of their roles in the resistance or their real social class position.

This form of class struggle facilitated the process described earlier. Fulfillment of 'quota' targets and satisfaction of the demand for land, even in terms of the decrees, were clearly aided by denouncing many rich, middle, and in some cases, poor peasants as 'landlords' or 'landlord agents'--the latter often by association within resistance organisations which 'landlords' supposedly had infiltrated. Membership of the resistance organisations had risen considerably during the 'united front' period. The VWP alone expanded from 20,000 to 700,000 members from 1946 to 1950.

> In effect, a united front was created *within* the VWP by the acceptance of members on the sole criterion of their willingness and effectiveness in fighting the French.[41]

When the policy changed and history was rewritten, 'landlords' were considered to have 'infiltrated' rather than 'joined' these organisations. The wide-ranging political struggle was thus symbiotically related to the need to appropriate land from those with per capita holdings in excess of the mean since the struggle both assisted and justified the appropriation. This relationship applied from the outset of Rent Reduction and Land Reform.

The Land Reform campaign in Hung Son village in Thai Nguyen province, the first village to undertake this campaign, revealed thirty 'landlords.' The later Rectification showed that only twelve of these were actually 'landlords' and the remainder were 'middle' peasants.[42] However, given the limited number of villages involved in the early waves, it may have been easier to exert countervailing pressure if the higher levels within the VWP wished to do so, than in the later, more extensive waves.

Admission of 'Errors' and Rectification

It has been argued above that Land Reform involved an attack on middle peasant interests because of the absolute land scarcity in the DRV; the land hunger of the poor and landless; the earlier distributions of land; the ambiguity of the classification; and, the method of implementation. These consequences were apparent during the early phases of both Rent Reduction and Land Reform and led to some modification of the relevant legislation. However, the VWP leadership's acceptance of the need for a comprehensive reform of the rural political and economic structure meant that it was not until 1956, after the completion of Land Reform, that major action was taken to deal with the adverse consequences. In addition, the absolute impact of the policies and deviations increased dramatically in 1955 and

1956, when the majority of villages underwent Rent Reduction and Land Reform--irrespective of whether the village-level impact had moderated or radicalised.

The cumulative impact--on the middle peasants and rural organisation, including the VWP--depressed the potential for output growth, although output inevitably rose in 1956 as a consequence of the end of the war. Corrective action was thus imperative, while the completion of Land Reform meant that recognition of 'errors' and their rectification could not prevent the implementation of a far-reaching political and economic reform. Changes in the policy and its implementation in 1954 may have made it subsequently more difficult to change the rural power structure. This line of reasoning is not meant to imply that the leadership could have stayed its hand earlier; indeed, its relative lack of control over the Land Reform organisations and cadres, coupled with and heightened by the existence of disagreements at the highest levels, did not permit it to do otherwise. Rather it argues that by 1956 the balance, in terms of objective conditions and within the VWP, had shifted in favour of action to mitigate some of the adverse consequences of Land Reform. In addition, developments stemming from Khrushchev's speech to the 20th Congress of the Communist Party of the Soviet Union in February 1956 provided ideological support for this action.

Admonition to avoid 'excesses' and respect the spirit and letter of the policies had been made throughout Rent Reduction and Land Reform, and was intensified in 1956 as the sheer numbers of villages and peasants involved increased. However, by the end of the final wave of Land Reform, it became clear that more decisive action by the central VWP leadership was required to deal with the serious consequences of the policy and its method of implementation. A meeting of the Central Committee of the Fatherland Front in July 1956 foreshadowed later developments. The statement was made that one of the tasks of the coming six months was 'to recapitulate and correct the mistakes committed in districts where Land Reform has been applied.'[43]

The Fatherland Front, the embodiment of the former 'united front' policy, was accorded significantly greater prominence during this period. It held a series of provincial conferences through mid-1956. Its Central Committee was in session when the communiques of the VWP CC and Council of Ministers were published at the end of October. The meeting at which General Giap made his speech was held under its auspices. Subsequently, members of the Front's Central Committee were sent to assist with the Rectification campaign.[44] Influenced by the Soviet Union and the '100 flowers' in China, the period from August to December 1956 was also marked by a policy liberalisation with respect to intellectuals.

In August 1956, in a letter to rural cadres on the completion of Land Reform, Ho Chi Minh referred to certain failings in the implementation of the policy, notably concerning holdings and yields, and reorganisation of the VWP, which required rectification.[45] In retrospect, this may be seen either as a further admonition to carry out the Land Reform correctly in the belief, or hope, that major corrective measures would not be required, or as preparing the ground for the public announcements later that year. Clearly a debate was continuing within the VWP leadership over the need for and extent of corrective action.

It was the September Plenum of the 2nd VWP CC, however, that brought about major public recognition of serious difficulties caused by Land Reform. The communique of this meeting, made public at the end of October after prolonged debate within the VWP, State, and army, confirmed that the problems manifest during the trial waves of both Rent Reduction and Land Reform had continued during the later waves,

> especially in the seventh and eighth drives of the movement for Rent Reduction and the fourth and fifth drives of Land Reform, local errors became generalised and aggravated.[46]

As argued above, the three crucial areas of concern were the classification, the reliance on poor peasant cadres, and the consequences for the VWP of the emergence of a parallel structure in the form of the Land Reform Committees.

> Criteria for distinguishing various classes in the rural areas were not well assimilated, the Party's policy of relying on poor and landless peasants was not well understood, land reform cadres wrongly valued our own organisation in the countryside.[47]

The publication of the VWP CC communique was accompanied by a series of meetings of the leading organisations in the DRV and speeches by the VWP leadership including, significantly, one by General Giap, the hero of the war against the French, to a Hanoi public meeting in which numerous examples of excesses and injustices were cited.[48] Giap referred explicitly to the insufficient attention being given to 'middle' peasants with the consequence of making them uncooperative, and also to the over-estimation of the number and significance of 'landlords', with the result that 'rich,' 'middle,' and even 'poor' peasants were incorrectly classified. He claimed that the three major 'errors' occurred in the application of the classification, the 'search for enemies,' and the organisation of the VWP.

Corrective action was also taken at this time with the announcement of the castration of powers of the central and local Land Reform committees, henceforth to become 'purely research organisations'; the abolition of the Special People's Tribunals; the disciplining of leading VWP and State officials; and, the replacement of Truong Chinh as VWP Secretary by Ho Chi Minh himself. Truong Chinh had been closely associated with the development and implementation of Land Reform and, although he retained his other positions and remained an authoritative and influential figure, his demotion had at least public significance.[49]

Further corrective measures involved amendments to the class classification, though these were minor; the release of those wrongly accused and imprisoned--estimated at 12,000 in most sources; and the rehabilitation of large numbers of expelled VWP members. A campaign of Rectification of Errors was subsequently launched and lasted until early 1958. The most difficult aspects of this campaign involved the restitution of lands and property, and the reorganisation of the VWP and other local organisations. In addition, abuses of the rights of religious and ethnic minority groups were to be corrected.[50] The Rectification campaign involved the popularisation of the VWP CC and Council of Ministers' resolutions and self-criticism, to be followed by reclassification and finally, by restitution and redistribution of land and other property.[51]

The October-November 1956 meetings and announcements served to publicise in dramatic fashion, the seriousness of the problems and their cumulative impact. At the same time, the new policy announcements and campaigns halted the activities of the Land Reform committees and attempted to reconstruct the VWP which had been severely affected in many areas. The virtual autonomy with which the Land Reform committees and cadres had acted had been a consequence of the desire for a thoroughgoing struggle against the rural rich on both economic and political levels. The 'excesses' had resulted from this in addition to the absolute land scarcity, the problem of neatly defining a small 'feudal' class, and the earlier redistribution of land.[52]

The necessity for corrective action reflected the impact of 'excesses' not merely on the VWP or individual 'landlords' and 'rich' peasants who were misclassified, but also on a substantial group of 'middle' peasants whose economic and political significance was considerable. They were economically important because of their status as surplus producers. Their political significance came from their position within resistance and patriotic organisations[53] in addition to the political strength derived from economic power. As argued earlier, the poor peasant demand for land, fuelled by the process of the Land Reform class struggle and confronted by the absolute land scarcity, was capable of producing a pronounced tendency to egalitarianism with severely adverse consequences for middle peasants. This tendency found justification in the ambiguities of the relevant legislation; application of the 'letter' of the Land Reform laws could violate the 'spirit' of the class line on which they were based--a class line rooted in both economic and political expediency.

Thus, although the speeches and resolutions of late 1956 dealt with 'excesses' and 'errors,' many of which had obviously occurred, the essence of the difficulties derived not from excesses--of either accident or design--but from the nature of Land Reform itself. It was the accumulation of middle peasant discontent which necessitated the admission of 'errors'. The injustices and ill-treatment of 'landlords' would not alone have justified such admission.

> It does not seem very likely that the injustices committed against Party members of landlord and rich peasant origin would have been adequate, by themselves, to stimulate a campaign for the correction of errors....But the number of middle and poor peasants who had been unjustly attacked and been large enough to make a correction of errors campaign obligatory.[54]

Although Giap and others blamed the 'errors' on the implementation of Land Reform, they were an inevitable result of a deliberate policy which was discredited by both the peasantry and the USSR after February 1956. The peasant opposition, resulting in uprisings in places, made action imperative at some stage, while the shift in the USSR legitimised a less-Stalinist (and less-Maoist) approach. It was this factor, coupled with the completion of Land Reform, that determined the timing of the admission of 'errors.'

The extent of the discontent in rural areas is illustrated by the extent of the reaction by the leading organs of VWP and State, in addition to the actual peasant uprising in Nghe An province in mid-November 1956.[55] A report issued after the meeting of 600 senior army cadres which lasted from 23 to 31 October 1956, illustrates the discontent felt in that quarter:

> The cadres expressed deep concern about the damage caused by the serious mistake in the application of land reform and the readjustment of organisations. At first, several of the delegates feared that the Central Committee and the General Military Committee were not fully cognisant of the situation in the countryside and had not rightly appraised past mistakes and achievements; they therefore warned the meeting against certain points in the resolution.[56]

It follows from the preceding analysis that although the activities of the Land Reform cadres could be halted, specific restitutions undertaken, and efforts made to respect the rights of particular groups, the underlying problem could not be solved. The basic fact remained that insufficient land existed to reconcile poor and landless peasant demands and aspirations with the maintenance of middle peasant holdings and support. The Rectification campaign involved a further attempt to reconcile these differences in the interest of economic and political stability, although the time involved in the campaign illustrates clearly that the admission of difficulties was far easier than their resolution.

Land Distribution and Landholdings

By the end of the agrarian reforms in the Delta and midlands in 1957, 810,000 hectares of land had been distributed with right of ownership to some two million peasant families. This represented 50% of the cultivated area in the 3,653 villages involved. In addition 100,000 animals and two million agricultural implements were also redistributed. As Table 2.1 showed, much of this land had already been acquired and temporarily allocated prior to the Land Reform campaign, during which a further 222,000 hectares were obtained. The significance of Land Reform, therefore, lay more in the consolidation of acquisitions and legalisation of redistribution than in the actual further acquisition of land.

Since Table 2.1 refers only to land owned

prior to 1945 or 1949 by groups whose holdings were to be appropriated during the agrarian reforms, it does reflect the fact that much of this land had been acquired by, or for, the peasantry before Land Reform. However, it does not reflect the amount of land distributed during Land Reform, since all land distributed to 'rich' and 'middle' peasants before Land Reform was liable to be reallocated during this campaign. This was the most complicated problem encountered during both Land Reform and Rectification, since it was here that the confrontation between middle and poor peasant interests was most acute.

Table 2.1 also gives no indication that land was requisitioned (or confiscated) from 'rich' peasants. This omission is undoubtedly of political significance since, not only was the requisitioning of 'rich' peasant land officially sanctioned by Land Reform legislation--where there was insufficient land obtainable from other sources--but also such requisitioning clearly did occur. Tables 2.2 and 2.3 demonstrated a fall in land owned by 'rich' peasants from 7.1% of the total in 1945 to 4.7% or 4.4% in 1953. There was also a fall in the population percentage from 2.9% to 1.6%--families from 2.2% to 1.2%. The land owned by this class fell further to 1.8% in 1956 (Table 3.5) without any reduction in the population or in the families' proportion. Only part of this loss of land by the 'rich' peasant class (from 113,000 hectares in 1945 to some 35,000 in 1956) would be accounted for by the reclassification as 'middle' peasants--with landholdings intact--as a result of making greater use of family labour.[57] Per capita holdings (Table 3.6) also show a fall in 'rich' peasant holdings.[58]

The extent to which land was taken from 'rich' and 'middle' peasants and given to 'poor' or landless peasants cannot be discerned precisely from the available data, nor can the extent to which such land was returned to 'middle' peasant use and ownership during Rectification. Given a hypothetical interpretation of the relevant laws, the extent to which this may have occurred has been indicated above (Table 3.4) on the basis of available data on land used by 'middle' peasants in 1953. There would also have been differences between various regions and villages depending on the following: the strength of the existing local power structure which might tend to favour rich and middle peasant interests; the moderate/radical balance within the local Land Reform Committee; and the local land scarcity situation.[59]

Available data for land used by 'middle' peasants in 1945--prior to temporary distribution of confiscated land documented in Table 2.1--show landholdings in excess of, or equal to, post-Land Reform owned land. This suggests that many of the (temporary) gains made by 'middle' peasants were subsequently put under pressure and reallocated. Table 3.6 shows 'middle' peasants using 1,688 square metres in 1945, and owning only 1,610 square metres in 1957. Demographic pressure would have reduced average holdings over the period since the total cultivated area in these villages remained roughly constant. Therefore, sufficient gains were retained to partially offset this pressure.

Table 2.3 showed 'middle' peasants using 46% of the land in 1953, whereas Table 3.7 shows them owning 42% in 1956--suggesting a loss of almost 10% during this period. The peasants would have held on to their land tenaciously as, in certain cases, they would have used the land for eleven years.[60] Furthermore, the general point that land used by 'middle' peasants would have been under threat during Land Reform obscures the considerable differences within this group which comprised some 36% of the rural population in 1953-56.

Table 3.6 below shows official data on pre- and post-Land Reform landholdings of the different groups within the rural population. It confirms the erosion of 'landlords' in both economic and political terms, and it demonstrates the extent to which the holdings of 'rich' peasants were reduced during the agrarian reforms.

Official data collected post-Rectification tend to show lower 'rich' peasant holdings than that collected immediately after Land Reform. This suggests that the appeasement of 'middle' peasant discontent was partially achieved by a

Table 3.6: Land Owned and Used Before and After Land Reform (in square metres per capita)

	1945 owned	1945 used[1]	1953 owned	1953 used[2]	1957 owned
Landlords	10,093	11,356	6,393	n.a.	738
Rich Peasants	3,975	4,679	3,345	n.a.	1,547
Middle Peasants	1,372	1,688	1,257	1,678	1,610
Poor Peasants	431	680	490	1,006	1,437
Landless Peasants	124	252	262	807	1,413

Notes: 1. Includes use of communal land.
2. Includes temporary redistributions including that of communal land.

Source: Tran Phuong (1968) pp.14 and 212, and Report of the Central Land Reform Committee 15/8/58, cited by Vo Nhan Tri (1967), p.201.

heightened attack on 'rich' peasants. It also suggests that a notable deficiency of the Land Reform legislation--namely, that land used by 'middle' peasants was to be appropriated (on the grounds that it was formerly landlord-owned) before land owned by 'rich' peasants--was altered or simply ignored during Rectification. The post-Land Reform data may also have been affected by cadres reporting 'rich' peasants as having large holdings in order to justify further attacks on them.

A number of major inconsistencies in the data reported on 'rich' peasants have already been raised. First, there was no explicit reference to land appropriated or confiscated from this strata, and second was the reduction in the proportion of the population in this strata. Of significance here is the fall in per capita landownership--over 50% according to Table 3.6, whereas other sources show no fall at all.[61] Since 'rich' peasants comprised barely 1% of all families and held under 2% of all land, such variations might be explained by sampling differences or variations in the criteria--in practice as opposed to in principle--for distinguishing 'rich' peasants from those subsequently referred to as 'upper-middle' peasants.

'Middle' peasants as a group are shown, by the official data in 1957, as owning approximately the same amount of land as used by 'middle' peasants in 1953. Land owned per capita by 'poor' peasants remained below that of 'middle' peasants, illustrating both the basis for egalitarian pressures and the fact that they were blocked. The data apply to 'middle' and 'poor' peasants as groups, and there would still have been considerable fluctuations in individual family circumstances. The impact at the upper end of the 'middle' peasant class would also have been greater. By 1957, the position of middle peasants, and particularly the richer members within this strata, was considerably enhanced in economic and political terms. The richer rural strata had been effectively destroyed.

In economic terms, the 1956 and 1957 rice yields of 1,800kg of paddy per hectare meant that some 1,330 square metres were required to produce the basic ration of 240kg of paddy per capita, or some 1,500 square metres to yield this standard after the payment of the agricultural tax.

'Landlords' were now at the level of agricultural workers prior to the agrarian reforms, while 'rich' peasants comprised a very small percentage of the population--1.7% in the Land Reform classification--and in 1956 held only 1.8% of the land. Hence, even if they possessed holdings sufficient to produce a surplus, in absolute terms it was still very small. Those 'middle' peasants with above-average holdings, however, had a clear capability for producing a surplus; the group as a whole held 42% of the land, exceeding the 41% held by 'poor' peasants. The 'midle' peasants with above-average holdings constituted some 12-15% of the rural population, and held 20-25% of the land (depending on the nature of the distribution within the class). Thus by 1957, they would have been of pivotal economic significance. This provided political strength which was increased by their contribution to the resistance and their membership in the VWP itself.

Table 3.7: Rural Population During Land Reform and Land Held, 1956

	% of rural population	% of rural families	% of land owned by class 1956
Landlords	2.5	2.1	1.1
Rich Peasants	1.7	1.2	1.8
Middle Peasants	36.2	31.9	41.7
Poor Peasants	42.7	44.1	40.5
Landless Peasants	13.0	16.7	12.0
Other Rural Workers	3.5	3.7	2.6
	99.6	99.7	99.7

Source: Tran Phuong (1968), p.162; Hanoi (1978), p.94.

In terms of the official classification the rural structure can be partitioned into two groups. The first comprised some 38% of the population and owned 43.5% of the land; the second group comprised 56% of the population and owned 52.5% of the land. This contrasts with the pre-1945 situation when the 'poor and landless' owned only 11% of the land, and the differences between 'rich' and 'middle' peasants were more clearly defined. A 'landlord' class comprising 4.0% of the population owned 25% of the land. However, the 1956-57 data obscure the differentiation existing at that time. Moreover, these two broad groups clearly overlapped. Those at the upper end of the 'poor' peasant class, possibly including cadres active in the VWP, would have held more land than those at the lower end of the 'middle' peasant class. It is likely that the distribution within both groups would have been relatively skewed in 1957.

The position of the poor and landless peasants had improved to the extent that they possessed per capita holdings capable, on average, of producing a basic level of consumption at existing levels of productivity. Hence Land Reform had substantially satisfied the peasantry's demand for land. A completely egalitarian outcome would have involved a major confrontation with middle peasants which was neither desirable nor feasible, given the crisis resulting from even a partial confrontation with this group. Crucially, it would have threatened to remove the surplus derived from the upper-middle peasants. Equal holdings would have resulted in a further augmentation of peasant consumption and, while in the medium term the potential of peasant family production could have been exploited via various price, incentive, and other policies, in the short-run, higher taxation of all peasants would have been required with attendant problems for the mass base of the revolution.

The new pattern of landholdings undoubtedly gave a significant boost to production during the years immediately following Land Reform. Cultivated area, per hectare yields, and production all rose, despite the further instability which occurred upon further classification and distribution during Rectification. However, during this period certain additional problems over the taxation of agriculture arose. Peasants found themselves liable for payments on land whose area and yield had been overestimated when it had been in 'landlord' hands during Land Reform.

The Rectification campaign was brought to a close in early 1958, some five years after the consolidated Rent Reduction decrees had been promulgated. Rectification itself had brought problems concerning classification, land restitution and reallocation, and within the local VWP and other organisations. Truong Chinh, in his report to the National Congress of the National United Front in March 1958, referred to the rehabilitation of unjustly treated 'poor' and 'middle' peasants as including some 'rich' peasants and 'landlords.' In some cases released cadres sought revenge, including the use of physical violence. Reconciling the conflicting groups was hence a difficult task. Truong Chinh also referred to only half the 'wrongly taken' land being returned along with varying proportions of other prop-

Table 3.8: Production During and After Land Reform, 1954-59

	1954	1955	1956	1957	1958	1959
Paddy acreage (m. hectares)	1.86	2.18	2.28	2.19	2.24	2.27
Paddy production (m.tons)	2.55	3.52	4.13	3.95	4.58	5.19
Paddy yield (quintals per ha.)	13.7	16.1	18.1	18.0	20.4	22.9
Total staple production (m.tons paddy equivalent)	2.83	3.94	4.79	4.39	4.96	5.77
Paddy per capita (kg)	190	256	292	271	304	335
Total staple per capita (kg)	211	286	337	301	331	372

Source: Official Vietnamese Data--see Appendix Table IV.

erty.[62] However, the question of what precisely constituted 'wrongly taken' land is a matter of conjecture. The difficulty in returning it confirms the problems detailed above, particularly the reallocation of land temporarily distributed before Land Reform. The regulations stated that it was necessary to ensure that the 'middle' peasant average remained above the 'poor and landless' average, while ensuring that the latter remained only a little below the overall village average.[63] The ambiguity of these guidelines and of the classification allowed for widely varying interpretations.

Rectification had to reconcile conflicting interests with respect to the scarce land available as well as within the VWP. To the extent that this was possible, it succeeded.[64] The conflict of interests between the peasantry and the rest of society, including the VWP, was more pervasive. At a political level, the leadership also faced problems in 1956. These problems included: the flowering of intellectual debate and dissent, itself related to Khrushchev's denunciation of Stalin at the 20th Congress of the CPSU and the launching of the '100 flowers' movement in China; the discontent of the Catholics in the DRV; and, the non-implementation of the Geneva accords which postponed reunification and caused discontent among those from the South who had regrouped in the DRV.

Land Reform had destroyed the economic and political base of landlordism and had transformed the rural structure to one no longer characterized by extreme differential access to the basic productive resource--land. The process by which this was achieved, however, contained a number of contradictions. Crucially, it confronted the peasantry's demand for land with the absolute land scarcity of the DRV, where securing a viable holding for all peasants would involve appropriating land from middle peasants. Second, it required the unleashing of 'class struggle' in contrast to the earlier pursuit of a class alliance to defeat the French. This meant relying on poor peasant cadres and, when taken to extremes, both by design --within the Central Land Reform Committee--and in practice, due to the lack of control of the leadership over the Land Reform teams and cadres, produced radical upheavals in all organisations including the VWP.[65] The reliance on these elements among the rural populace itself fuelled the demand for land which threatened the middle peasant economic and political position, and a major source of the surplus production on which the State depended.

Rectification became necessary when the balance between class line and class struggle was so tilted in favour of the latter that it threatened to overwhelm not only the VWP, but, more crucially, a major productive base of the DRV itself. Thus the role of Rectification was more than just to correct the damage done to the VWP; essentially it was to halt the confrontation with the middle peasantry who were not only numerically significant--over 30% of the population--but of crucial productive significance.[66]

The middle peasantry had little direct economic interest in Land Reform for it threatened some of the improvements in their landholdings over the previous years. Carried to extremes, it clearly conflicted with their material interests. The admission of 'errors' and Rectification were necessary to preserve their position and support, at least in the short-term, for even at this stage the preservation of a family-based peasant economy did not appear to be an objective for the medium- or long-term.

The Land Reform had achieved an essential political objective--the destruction of the landlord class. In so doing, it had liberated the economic potential of the peasantry to produce significant short-term productive gains. In the medium-term, it constituted a first step towards the collectivisation of agriculture, the logic of which is analysed in the following chapter.

Summary of the Land Reform Period, 1945-57

The previous two chapters have described and analysed the process of Land Reform in the DRV in terms of the developments of policy and its implementation.

The original 'land to the tiller' policy (following Comintern orthodoxy) was temporarily put aside during the nationalist resistance struggle against the Japanese and the French. This was because the struggle required a united front, and the landlords and rich peasants, whose land would have to be taken in any distributionalist Land Reform, were important both for their influence and authority and because they provided the surplus production which fed the Viet Minh forces.

It was argued that the balance between needing to appease this rich strata and fulfilling the demands for land coming from the majority of the peasantry began to shift in 1950-51. The reasons for this were the widening geographic, and hence, revenue base of the DRV; the consolidation of communist rule in China; and, the pressure from the poor peasantry who provided the bulk of the fighting forces.

During the first period--up to 1950--some reduction of rents and interest rates occurred, and some land was redistributed--mainly communal, French, and that belonging to absent or 'traitorous' landlords. It was argued that this temporary distribution of land was a contributing factor to the problems encountered during the actual Land Reform some years later.

The period 1950-53 was seen as an interim period between the earlier united-front line and the later anti-landlord and, in effect, anti-rich line. There was a gradual de facto hardening of policy with the actual implementation of earlier decrees. The new tax system, introduced in 1951, became the major source of government income and helped consolidate its position.

Despite the shift in position and consolidation of VWP power, it was not until 1953 that radical distributionalist Land Reform was implemented. The rationale for the delay, which was validated by later events, was that a united front was still essential and that Land Reform inevitably involved breaking this. Nevertheless, there were those who argued all along that mass peasant support was more important and required as well, the actual honouring of the land promise.

Additional reasons for the postponement of Land Reform were the relative cadre shortage, the absence of sufficient control over the rural areas, and a preoccupation with the military struggle against the French. Peasant support, or at least non-opposition, was guaranteed by the promise of a future distributionalist Land Reform.

By 1953 the VWP was in a stronger position, the military position in the war against the French had improved, the Chinese Land Reform was complete, the VWP had been partially purged, more cadres were available, and the resource and revenue base was stronger and wider. At the same time, pressure from the poor peasantry was increasing, no more land remained to be distributed (apart from that held by ordinary landlords and peasants), and an alternative regime existed to draw the support of at least the richer peasants.

These forces, along with what has been referred to as the 'weight of ideology' were pushing the DRV towards a radical distributionalist Land Reform.

The data relating to landholdings have been discussed at length to demonstrate the inherent problems of implementing a Land Reform of the type intended, in a situation where there was an absolute land scarcity and very few large landowners. A purely egalitarian Land Reform would involve the expropriation of the sizeable group of peasants with landholdings slightly in excess of the mean. This analysis was related to the actual Land Reform legislation and process of implementation, and revealed three inevitable areas of difficulty. First, the classification contained areas of ambiguity which were compounded by the attempt to fix class status in a situation of continual change and socio-economic mobility[67] and the tendency to overestimate the proportion of rural rich.

Second, the problem posed by land scarcity was compounded by some ten years of temporary land redistribution during the final Land Reform campaign. The data show that much of the land was in the control of middle peasants who would attempt to hold on to it tenaciously since, even with it, their holdings were still barely adequate.

The combination of these two factors made confrontation with a sizeable proportion of the middle peasantry inevitable, while the method of 'political struggle' sanctioned and exacerbated this confrontation. Although deviations from the letter and spirit of the legislation were criticised, Chapter III showed that the legislation was (necessarily) ambiguous, while, as in the case of China, it was better to be 'leftist' than 'rightist.'

Exaggeration of the numbers of rural rich, necessary in order to obtain sufficient land to redistribute, meant attacking middle peasants. The redistribution of land previously temporarily allocated, meant dispossessing many who had acquired land in previous years. The necessity of reliance on the poor and landless, and those 'untainted' by association with the rural rich, meant attacking existing rural and resistance organisations--including the VWP itself in places.

Chapter III, therefore, shows that 'errors' were not the results of mistakes and misinterpretations, as in most standard accounts, but were inevitable. The decision to act on these 'errors' in the form of a Rectification Campaign was, however, the result of additional factors--whether any of them alone would have been sufficient is unclear. The combination of crises in the lower levels of the VWP (which was to remain relatively weak in rural areas in subsequent years), middle peasant opposition with its impact on production and State procurement--both of which had an accumulative impact by 1956--along with the changed ideological environment resulting from Khrushchev's speech to the 20th Congress of the Communist Party of the Soviet Union and its impact throughout the Communist world, created a situation in which action to correct certain effects of the Land Reform was desirable, justifiable, and possible.

Rectification, however, faced the same contradictions inherent in Land Reform and only partially resolved them. The substantial group of peasants producing a small surplus was appeased and the situation in the rural areas stabilised, though, as analysis of the data shows, possibly at the expense of some rich peasants.

The preceding chapters have thus shown the limited ability of the VWP to impose its will on the peasantry. The next attempt came with collectivisation.

CHAPTER IV

Launching Collectivisation

Introduction

In 1957 the agricultural sector in the DRV was characterised by a mass of moderately differentiated peasant small-holdings. By 1960 it consisted of some 40,000 agricultural producer cooperatives to which over 85 percent of the peasantry belonged. This chapter analyses the origin of these cooperatives and the rationale for their development.

Land Reform had provided all rural households with between 0.5 and 1 hectare and had freed them from existing rental and debt obligations. This new structure and the recovery from the chaos of war and agrarian reforms produced a surge in agricultural production, culminating in the 1959 harvest, which was not surpassed for fifteen years.[1] However, in addition to the problems dealt with in the Rectification campaign, the new structure posed problems for State mobilisation of agricultural production.

The admission of 'errors' had shaken confidence in the VWP, while Rectification required reconciliation of local cadres who were less than 2 percent of the population in rural areas.[2] The extent of State and VWP control over the peasantry was thus limited. In addition to the appearance of intellectual and religious dissent, there was also a recognition that reunification of Vietnam would be postponed indefinitely or at least treated as such for planning purposes. This had economic implications, for the south of Vietnam had previously provided a food surplus for the deficit north.[3] The peasantry's independence of the State had increased while the State's dependence on the peasantry had risen--and would increase further if a policy of rapid industrialisation were to be pursued.

Industrial development was considered necessary to rebuild industrial capacity and provide a basis for economic development, partly by assisting agricultural development. In addition, industrialisation was seen as an inherent feature of a socialist economy, a view that was held equally strongly twenty years later.

> The only socialist economic foundation that can be achieved is extensive mechanised industrialisation. He who forgets that is not a Communist.[4]

The ideology of the VWP thus demanded industrialisation which in a mainly agrarian economy required resources from agriculture or from overseas. This ideology viewed cooperatives as a concomitant feature of socialist development in terms of their ability to fulfill the demands of industry. They were also considered a 'progressive' form of agricultural organisation capable of advancing production and the technical state of agriculture, as well as limiting differentiation and producing a more egalitarian distribution of income.

Three themes dominated all discussion of collectivisation and provided its rationale. First was the view of cooperatives as an inevitable progression from individual peasant holdings and an inherent feature of the transition to socialism. Small-scale peasant agriculture, it was argued, would lead to differentiation and the marginalisation of groups within the rural population and inexorably to the establishment of capitalist agriculture. Second, cooperatives were considered the only means for developing an extremely backward agriculture. In the short and medium term they would facilitate a more sophisticated division of labour and derive benefits from larger-scale organisation of resources. Third, cooperatives were viewed as a means for mobilising the agricultural resources required by the nascent industrial sector.

These themes are interrelated but it is important to identify them in order to avoid confusion when examining arguments and evidence cited in support of collectivisation, or against the smallholder agriculture which existed in the DRV from 1956 to 1959 and continued to be practised in subsequent years by those peasants outside the cooperative and, to a degree, on plots within the cooperative.[5] Other issues such as the need to develop a 'socialist consciousness', the necessity of political integration,[6] and the influence of contemporary Chinese experience will be considered within the framework of these three themes.

The DRV's Own 'Scissors Crisis'

Table 3.8 showed that rice production rose in 1956 by 17% and total production of staple food-stuffs by 22%. The two favourable harvests of that year increased the effective demand from the peasantry for consumption goods and production goods such as fertilisers and tools. Demand was also stimulated by the distribution of property and income during Land Reform which had produced a more egalitarian income distribution and hence a higher average propensity to consume.

This increased demand for consumer goods, which was further boosted by higher government expenditure financed largely by increases in the money supply, encountered constraints on the supply side. Imports were limited, production re-mained low in the manufacturing sector, and most reserves of consumer goods remaining from the French period had been exhausted. Imports of con-sumer goods fell in value terms by 61% between 1955 and 1959. As a proportion of total imports, they fell from 55% in 1955 to 32% in 1956, rising marginally again in 1957 to 33% (possibly in response to the supply problems) before falling to 25% in 1958, 16% in 1959, and 13% in 1960.[7] The supply and demand imbalance caused a shift in the relative prices of agricultural and non-agricul-tural commodities which the State could do little to prevent given its limited influence within wholesale and retail trade (shown in Table 4.1).

In 1957 the terms of trade moved sharply against agriculture as the market adjusted and speculators took advantage of the growing short-ages of consumer goods. This shift which began in the second half of 1956 had a disincentive effect on the peasantry who limited their sales of agri-cultural products, and instead used the production increases to raise their own consumption and build up reserves.

Changes in the quantity purchased accounted for

Table 4.1: State Share of Wholesale and Retail Trade, 1954-60 (percentages)

	1954	1955	1956	1957	1958	1959	1960
Wholesale trade	..	28	45	53	74	89	94
Retail trade	10	20	36	40	58	80	91

NB: In respect of retail trade, the State sector includes supply and marketing cooperatives, the mixed 'State-capitalist' sector, trading cooperatives, as well as State retail stores.

Source: Nguyen Tien Hung (1977), p.63 citing official DRV sources. Also, Le Chau (1966), p.260, Nguyen Xuan Lai (1957), p.37, and Vo Nhan Tri (1967), pp.239, 243, 331, 332 and 336.

the major part of the variation in the State share of paddy production (shown in Table 4.3). There were, however, also problems with tax collection,[8] partly resulting from the overestimation of areas and yields during Land Reform. These had remained a problem despite efforts to adjust them during Rectification.[9]

In 1957 food production fell mainly because of a decrease in the area of 5th month paddy and the effect of drought, on subsidiary crops in particular. The disruptions caused by the Rectification campaign and the attendant reclassifications, redistributions, and reconciliations may also have contributed to the 8% fall. However, the price 'scissors' were at their widest during the first half of 1957 due to the lagged effect of the supply and demand factors exacerbated by speculative action, especially in textiles. Inflation at this time was equivalent to 20% per annum. Peasant discontent with the deterioration in the terms of trade would thus have been greatest during the period of harvesting a relatively poor crop. State concern at the possible impact of this on procurement, and hence potentially on food prices and the urban workforce, is illustrated by the decree which was issued in April 1957 pro-

Table 4.2: Price Movements, 1955-59--Indexes

	1955	1956	1957	1958	1959
All consumer goods	100	113 (Dec.)	125 (May)	....	....
Purchase price of agricultural commodities					
-foodstuffs		100	96.5	97.0	97.5
-livestock products		100	107.4	101.8	101.4
-industrial crops		100	100.4	87.8	87.0
Index of 10 basic necessities		100	109.4	97.2	96.3

Source: Official sources cited in Le Chau (1966), pp.207-8.

hibiting speculation, particularly in textiles and pork.[10]

Prices of livestock products rose in 1957 by almost as much as the '10 necessities' index, suggesting that the peasantry indicated its opposition to the terms of trade by withholding meat supplies. The price of meat remained high in subsequent years. The increased industrial investment, rising urban income, and high income elasticity of demand for meat were major contributory factors. Meat procurement continued to be a problem for the State.

The action by the State to prohibit speculation, to restructure the prices of industrial commodities, and to control an increasing proportion of wholesale and retail trade apparently had the desired effect in the latter part of 1957 as prices fell and sales of produce by the peasantry increased.[11] In price terms the crucial sales were those of non-rice crops, although State action was also necessary to ensure that reduced sales of rice did not lead to increased rice prices and serious consequences in terms of the response of the urban population. Although most of the difficulties were blamed on speculators, their action was in response to the fundamental supply constraints and limited State control of prices and procurement.

The State share of the paddy crop in 1957 was below that of 1955 and the targeted level, although price movements had been brought under control. In 1958 in spite of the fact that the terms of trade had returned to their 1956 level, procurement difficulties continued, indicating that State control of price alone was not sufficient. The 1958 5th month paddy harvest was extremely poor with the same area as in 1957 and with yields down to the 1955 level. In terms of the good 1956 harvest, production was down 24%. As a result of this and the continuation of the problems described above, food procurements in the first half of 1958 were disastrously low--only 20% of the planned level and much lower in certain provinces.[12]

The peasants were maintaining increases in their own consumption--achieved as a result of the previous good harvests and the abolition of landlordism--and building reserves against future food shortages. With respect to meat products, they took maximum advantage of their ability to trade independently. The terms of trade and shortage of consumer goods limited the incentive to forego consumption of their own production.

<u>Table 4.3: State Share of Paddy Crop (Taxes and Purchases), 1955-60</u>

<u>Year</u>	<u>Paddy (tons)</u>	As % of total <u>paddy production</u>	
1955	681,700	19.4	
1956	534,000	13.0}	
1957	590,400	15.0}	tax alone 9.0
1958	674,000	14.7}	
1959	852,700	16.4	
1960	782,000	18.7	
1960 All Staples	878,000	18.7	

<u>Sources:</u> Official figures cited by Vo Nhan Tri (1967), pp.245 and 337, Le Chau (1966), p.210 and Hanoi (1977c), p.108.

The bumper 10th month paddy harvest of 1958 alleviated the problems of the first half of the year although the State percentage for the whole year remained less than 15% of paddy production. The situation continued to improve in 1959 as production rose to record levels for both the 5th and 10th month paddy harvests,[13] prices stabilised, and the terms of trade shifted in favour of agriculture. By 1959 the State sector controlled 89% of wholesale and 80% of retail trade. Industrial production had also begun to expand rapidly. However, the peasantry's tendency to raise its own consumption of rice and increase reserves is shown by the limited increase in the State share of production and the apparently high marginal propensity to consume--of the 610,000 tons extra paddy produced in 1959 (over 1958) the State obtained only 178,000 or 29%.

This limited willingness to trade--which may have been more marked for other products--was the result of several factors: the historically low levels of consumption and hence the substantial room for improvement; the relative terms of trade which, although better than in previous years, may well have been considered unfavourable by the peasantry; and the continuing shortage of consumption and production goods such as fertiliser and tools.

Table 4.4: Paddy Production by Harvest, 1955-60 (million tons)

	1955	1956	1957	1958	1959	1960
5th month	1.16	1.58	1.45	1.20	1.80	1.16
10th month	2.44	2.55	2.50	3.36	3.40	2.92

Source: Karamyshev (1959), p.33, Hanoi (1977c), p.121.

In 1959, 3.72 million tons of paddy or the equivalent were required to provide a ration of 240kg per head for the whole population. The excess was therefore 1.48 million tons in paddy (over 2 million tons in all staples) of which the State obtained in tax and purchases, 0.85 million tons or 58%. In 1960, with 86% of the peasantry in cooperatives and with a lower harvest, the notional excess was only 0.34 million tons of paddy although the State obtained 0.78 million tons. The notional excess of all staples in 1960 was 0.86 million tons; the State obtained 0.88 million tons.

In June 1958 a Council of Ministers' communique on tax collection and procurement noted the need to 'mobilise the masses in the study of the important question of food supply.'[14] The price and procurement problems of these years showed that the State required control not only over markets and trade to ensure that supply constraints did not result in inflation, distorted terms of trade, a reluctance to trade on the part of the peasantry, and their attendant consequences, but that it required more direct control if procurement targets were to be met. Land Reform had made the peasants independent not only of landlords and usurers, but to a large extent, of the VWP and State as well.

Regardless of the implications of an independent peasantry for production, the immediate perceived need of the VWP leadership was for rapid industrialisation and an extraction of resources from agriculture which an independent peasantry might not willingly provide. As a later study contended:

> Land Reform constituted the first step, freeing the peasants from an untold misery....But it is really impossible to draw out of some million individual exploitations, each covering hardly half a hectare, a notable volume of marketable agricultural products to supply to towns and industrial centres in full development, and to export in exchange for equipment.[15]

The production and procurement experience of these years provided the background for the early development of forms of cooperative organisation within agriculture. Another aspect of the rural situation which received attention during these years was the extent of, and alleged increases in, the degree of differentiation among the peasantry. It is possible to distinguish different material interests among sections of the peasantry in relation to the price and procurement issue.

Peasant post-Land Reform landholdings were capable on average of producing basic food consumption requirements. In the relatively prosperous years of the late 1950s, the majority of peasant producers would have been able to meet family consumption requirements in addition to paying land taxes and trading for other basic necessities.

Increases in the prices of non-agricultural commodities 'essential' in peasant consumption (i.e. with low price elasticity of demand) would have resulted in increased marketing of agricultural output in order to maintain purchases (assuming goods were actually available). Increases in prices of other non-agricultural commodities (with high price elasticity of demand) which would form a larger part of richer peasant consumption would produce a fall in marketed output of agricultural commodities. In the case of agricultural commodities produced solely for exchange, resources would be transferred to the production of consumption commodities including non-food items. Commodities produced for use or exchange would be increasingly consumed within the family. The major part of these 'discretionary' marketings would have come from upper-middle peasants which illustrates the importance of this group which constituted probably 15 percent of the peasantry in these years.[16]

Poor peasants with inadequate resources may have been affected particularly severely by these price movements. They would have received low prices for rice and other basic staples while paying high prices for other 'necessities.' Thus their interest in cooperative forms of agrarian organisations and greater control of markets and prices may have been stimulated during the period. The differing material interests of groups within the peasantry and the differing conceptions of them provided further background for the development of cooperative forms of agricultural organisation.

First Steps--Mutual Aid Teams

A form of cooperation in agriculture had always existed in Vietnam. It provided a basis for the development of Mutual Aid Teams (MATs) in which the peasants--while retaining their own private property--assisted with various production tasks. These teams had apparently been formed in Thanh Hoa province as early as 1946.[17] Building on the traditional forms, these early MATs were organised for seasonal tasks only and included a handful of families.

Development of MATs on a large scale began during Land Reform in regions where that campaign had been completed. By August 1955, 31,657 MATs had been formed in 447 villages representing some 10 percent or more of peasant families.[18] By December 1955, 67,000 groups had been formed and by the beginning of 1956 this number had risen to 153,000 teams. The majority of these were temporary groups operating on a seasonal basis and consisting of seven families on average. The objective was for them to become permanent teams as a first step towards the development of cooperatives.[19]

In March 1956 a national Mutual Aid Conference called for the development of MATs as the main force for promoting agricultural production.[20] The conference set targets for the inclusion of 40% of peasant families in MATs by the end of 1956 with 15% of them in permanent teams. The targets for areas where Land Reform had been completed for some time were, respectively, 80% and 22%.[21] These targets were later raised and by the end of June 1956, 185,700 MATs had been formed including 59% of peasant families. The overall objective, incorporated into the final version of the 1956 State plan,[22] was for at least 60% incorporation by the end of 1956. The peak of the first major development of mutual aid was over 190,000 MATs, including more than half the peasantry. That level was not exceeded for another two years.

The Rectification period seriously retarded the development of MATs as well as other forms of rural cooperation. Credit cooperatives had been established in areas populated by over half of the peasantry, but they had only 143,000 member families. Funds mobilised and loaned represented five dong per member family on average, equivalent to 10kg of rice, and suggested an unequal access to credit at this time.[23] Supply and marketing cooperatives had less than one million members. The formation of experimental producer cooperatives was at an embryonic stage and hence not affected directly, although it was retarded indirectly because of the effects on MATs.

At the end of 1956 there were sixteen 'low level' cooperatives in which land was individually owned but collectively farmed, with members receiving both labour and rental income, and only one 'high level' cooperative in which land was collectively owned. During Rectification many MATs disbanded or became inactive. By mid-1957 approximately 72,000 were operating with any degree of regularity.[24]

As Rectification proceeded, attention focussed on continuing the momentum of institutional change. A conference in May 1957, addressed by both Ho Chi Minh and Truong Chinh, laid down as the central task in the countryside, the development and consolidation of MATs and their combination to form cooperatives.[25]

A major drive to develop and encourage MATs occurred during the 1957 10th month harvest. It was linked to the re-education of rural cadres and VWP consolidation which were also occurring at this time. Emphasis was placed on the potential of MATs in terms of organisation and of their claimed higher yields. This drive followed a September decree of the VWP on the restoration and development of MATs.[26] By the end of the harvest there were over 85,000 teams--including around 25% of peasant families--of which 17,000 were permanent teams.[27]

In 1958 there was a steady expansion of MATs so that the total reached 157,000--41% of peasant families by July. The end of the year target was 60% in permanent teams.[28] With the completion of the Rectification, MATs became the priority with promotional tours by Ho Chi Minh, Truong Chinh, and Le Duan taking place in many regions. By the end of September there were some 195,000 MATs--more than 50% of peasant families.[29] A further drive was launched during the 10th month harvest. There was a greater emphasis on the productive importance of MATs at this time, whereas later and more generally between harvest periods, their role in 'socialist transformation' was stressed.

Table 4.5: Peasant Families in Agricultural Organisations, 1955-60

Year	Individual Production	Mutual Aid Teams Seasonal	Mutual Aid Teams Permanent	Cooperatives Low Level	Cooperatives High Level
1955	59.5	37.2	3.3	---	---
1956	49.9	45.1	5.0	---	---
1957	78.1	19.9	2.0	---	---
1958	29.6	44.7	21.0	4.7	---
1959	17.1	26.2	11.3	43.0	2.4
1960	<--------14.2[1]--------->			73.4	12.4

Note: 1. No data is available on remaining MATs in 1960.

Source: Derived from data in Hanoi (1960e), pp.84-86, Hanoi (1978), p.98.

Table 4.5 shows the shift from independent peasant production to MATs in 1958 and to low-level cooperatives in 1959 and 1960. The subsequent, less dramatic shift to high-level cooperatives is considered in Chapter V. Theoretically MATs can be perceived in two ways. First, they could serve as an aid to individual production in overcoming seasonal labour shortages and facilitating joint ownership of indivisible production inputs such as a buffalo. In conjunction with supply and marketing and credit cooperatives, they could augment a rural economy based on an independent peasantry. Second, they could constitute a transitional form of organisation providing a basis for producer cooperatives. The progression envisaged in the DRV was from temporary MATs--involving a pooling of labour for seasonal tasks--to permanent MATs in which earnings were distributed according to labour input valued on a points basis. The supply and marketing cooperatives, although organised separately from the Internal Trade Ministry, were a means for increasing State influence over the supplies to agriculture and the marketing of its produce. Credit cooperatives were a way of mobilising peasant savings and channeling loans into areas (e.g. the producer cooperatives) favoured by the State.

Peasant participation in forms of mutual assistance would be expected to fluctuate depending on market conditions. However, the rapid withdrawal of considerable numbers from MATs in 1957 suggests that their existence was more apparent than real, and that peasants were unconvinced of their advantages. It also accords with the later admission that coercion had been employed in their formation: 'the mobilisation of the masses to take part in mutual aid teams usually had the character of pressure.'[30]

The price and procurement difficulties of these years, the poor 5th month harvest of 1958, and the earlier example of collectivisation in China produced a concerted drive in the latter part of 1958 to establish permanent MATs as a transition to the development of cooperatives. As Table 4.5 shows, some 50 percent of the peasantry moved from individual production to some form of cooperation in 1958.

The communique of the September 1958 Council of Ministers' meeting made clear the objectives for the 10th month harvest period now that Rectification was complete:

> It is necessary to praise and emphasise the important and decisive role of Mutual Aid Teams and agricultural cooperatives in agricultural production. The Council of Ministers is unanimously determined to consolidate and develop Mutual Aid Teams. For the time being, our main task is to set up permanent Mutual Aid Teams and to practice the 'common discussion and evaluation of work' method--and set up agricultural cooperatives in the regions where

> conditions allow...so that we will be able to solve all problems--the most important of which is the application of new agricultural techniques, thus ensuring increased agricultural production.
>
> The Council of Ministers stresses the close and uninterrupted relations between the grouping of peasants and the settlement of all problems relating to agricultural production, irrigation, fertiliser, cattle, agricultural implements and seeds, and the application of new agricultural techniques. If we do not group peasants into Mutual Aid Teams and cooperatives, we will not be able to solve these problems and apply new agricultural techniques to increase production and raise output. These are our experiences, and those of China in its present leap forward.[31]

Chinese experience had influenced Land Reform in the DRV and it continued to be significant in collectivisation.[32] In mid-1958 a series of articles on Chinese experience was published in *Nhan Dan* and numerous references were made to it by VWP leaders.[33]

The major drive to create MATs thus occurred during the 10th month harvest of 1958. By the end of the year nearly two million peasant families (over 65 percent of the total) were in MATs. Those outside of MATs included the population of the mountainous areas then experiencing Land Reform and the upper-middle peasant groupings. This was the peak of MATs for they were rapidly transcended by the development of cooperatives, as was made clear at the September 1958 Council of Ministers' meeting and further stressed at the 14th Plenum of the 2nd VWP CC in November 1958.

Collectivisation

The 14th Plenum was followed by the presentation of the 1958-60 Three-Year Plan and in December 1958 by a major conference on agricultural cooperation. These set over-optimistic production targets and the objective that 'under the Three-

Year Plan, ending in 1960, almost all peasant families will have joined elementary agricultural cooperatives.'[34] As an indication of the absurdly exaggerated claims which were current during this period, undoubtedly heavily influenced by China, the declared objective for the 1959 5th month harvest was to exceed the 1958 10th month harvest. The normal division of total production was 40% for the 5th month and 60% for the 10th month. Moreover, the 1958 10th month harvest had been unprecedently high; with the exception of that of 1959, no 10th month harvest (during the period to 1975) exceeded the 1958 level, let alone any 5th month harvest. The paddy production target for 1960 was 7.6 million tons, 45% greater than the record 1959 harvest, and still 37% higher than the record 1974 crop.[35]

The 16th Plenum of the 2nd VWP CC in April 1959 and the major speech by Truong Chinh to the National Assembly in May 1959 provided further details of the collectivisation programme.[36] When compared with the late 1958 resolutions, these place greater stress on 'consolidation', given the progress made during the intervening months. They also emphasise the alleged negative aspects of private production as an argument for collectivisation. The three themes of collectivisation as a means for developing techniques of production, ensuring procurement, and reducing differentiation dominate all policy statements of this and subsequent periods.

The Council of Ministers' resolution cited above stressed the relationship between cooperative forms of organisation and the development of production techniques. The argument was two-fold. In the absence of industrialisation and mechanised inputs to agriculture, the collective use of land and labour could benefit from a more rational layout of plots, a more sophisticated division of labour, and the organisation of labour on a larger scale for irrigation projects. Secondly, larger units would make it easier to introduce new inputs and take maximum advantage of the economies allegedly inherent in them.

> Collectivisation is also the basis for our bringing about a technical revolu-

tion in agriculture and advancing to the mechanisation of the whole branch of agriculture.[37]

The list of new or developed techniques which, it was claimed, could be more easily applied in co-operatives included: irrigation, fertiliser--both organic and chemical, improved seeds, deep ploughing, close planting, pesticides, destruction of worms, improved tools and machinery.

The experiences of China and even the USSR were cited to support the contention that collective working could both facilitate the introduction of new techniques and substitute for them.

> Practice has clearly shown that: with the change of the relations of production, with cooperation in labour, thanks to the patriotic emulation movement which provides a strong stimulus, our peasants are not only able to increase production but can also apply the advanced experiences of the brother countries in a creative manner, particularly the experiences of China in improving farming techniques.[38]

In alluding to Chinese experience, this argument that cooperatives would result in higher production and productivity through the application of new techniques and better utilisation of existing techniques, merged with the belief that cooperatives were appropriate to socialism. 'A country which claims to be Marxist-Leninist must of necessity build socialism with a collectivised agriculture.'[39]

The influence of China should not be overemphasised. The Chinese always considered cooperatives as transitional, to be superseded by communes, while the VWP always called cooperatives 'socialist.' The VWP was critical of the 'Great Leap' as well as communes, and believed significant output increases could be achieved by cooperatives. The converse of this argument was that individual peasant production was dependent on backward techniques practiced on plots suffering diseconomies and that reliance on it would lead to agricultural stagnation with consequences for the rate of growth of industry and the economy. It was argued that by 1959 the potentiality of individual peasant production had been realised.[40]

The subsequent development of agriculture in the DRV, the increasing awareness that mechanisation--or even semi-mechanisation--was a distant prospect, and that the only sources of productivity improvements would be labour-intensive developments and the widespread application of existing best-practice techniques and improved organisation are discussed in Chapter V. The argument employed in 1958-59 and subsequently was not based on the availability of mechanised inputs, but stressed that cooperatives were viable even in their absence:

> In industrialised countries, agricultural cooperativisation rests essentially on the mechanisation of agriculture. In the DRV it was carried out before industry could supply machines to peasants: that was a major difficulty. However, even when the working tools and methods of working did not undergo notable changes, a new productive force which is the collective force, was brought into being through cooperativisation.[41]

The weakness of the argument that cooperatives were the key to the development of the forces of production lies in the notion of increasing returns to scale in agricultural production, particularly in wet-rice cultivation. Almost all of the production inputs referred to--such as fertiliser, improved seeds, and irrigation--can be applied on micro-plots without decreasing returns. The possible exception is large machinery, the widespread use of which was unlikely in the foreseeable future in the DRV. Subsequent experience showed that cooperatives suffered decreasing returns with respect to organisation and management. They required skilled cadres and technicians of whom there was an acute shortage, and they had difficulty motivating the peasantry to believe that cooperatives served their material interests. There were also problems ensuring that cooperatives served the interests of the State.

The increase in production up to 1959--far from being the best result achievable from 'petty agricultural economy'--was the best achieved for fifteen years. The lessons provided by the years from 1956 to 1959 may well have been a better indicator for agricultural performance and potential than references to Soviet Kolkhoz and Chinese collectives. As Rene Dumont pointed out:

> The Land Reform, despite its errors, was accompanied and immediately followed by a large increase in production. <u>The establishment of cooperatives marks on the contrary a distinct slowing down of the rate of increase.</u>[42]

Despite the lack of evidence to support this argument in the DRV or elsewhere, it was frequently asserted that Land Reform did not ensure the conditions for a development of productive forces and the application of new technologies.

> To march forward we ought to reform small agricultural production based on individual exploitation and intensely develop the agricultural productive forces. Agricultural cooperativisation thus became a necessity, the sine qua non for technical...progress in the countryside. Either we succeeded in it, or we would be compelled to make a step backward in agriculture or at best to stagnate.[43]

The case for cooperatives, however, did not rest solely on their alleged greater productivity and ability to apply and develop new techniques. Indeed, this appears to be the weakest of the three rationalisations. The second theme which permeated discussion during this period was the alleged negative consequences of independent peasant agriculture, not in overall production terms, but for intra-peasantry differentials. From this idea arose the contention that it would lead inexorably to the emergence of capitalist agriculture.

We must bear in mind that the system of

private ownership of the means of production and individual production still provide the material basis and social conditions for the appearance of the spontaneous tendency to capitalist development.[44]

This argument, stressed by Truong Chinh in particular, had a number of elements. The first was that differentiation had increased since Land Reform and that if it continued, land would become concentrated in the hands of a small minority of the peasants, a class of agricultural labourers would develop, and the rural class structure would become highly differentiated. This would lead to resentment within the emergent poor peasantry and opposition to the VWP, thus endangering the worker-peasant alliance.' A powerful, rich peasant grouping, independent of VWP and State control, would be capable of withholding produce, interfering with markets, and engendering capitalism in other sectors.

Two main causes of differentiation were cited. The unequal ownership of land and other property which characterised the post-Land Reform rural structure gave certain peasants initial advantages. In addition, the size of individual plots, averaging little more than half a hectare per family, meant that natural or personal calamities could easily result in debt and loss or property. The frequency of these calamities and the potential which still existed for usury and trade speculation would accelerate this process.[45]

Some 'evidence' on differentiation is considered in the following tables. Table 4.6 is derived from data normally cited to show that 'rich' and 'middle' peasants had a higher average living standard than 'poor' peasants and 'landlords' on the grounds that the former categories contained proportionately more surplus producers than the latter categories.[46] Family classification was that which had been fixed during Land Reform. This interpretation fails to take account of the fact that 60 percent of rural families were in the 'poor' peasant category, and just over one percent in the 'rich' peasant category.

By weighting according to these proportions, Table 4.6 shows the inadequacy by 1959 of the Land Reform classifications. It also shows that while former 'poor' peasants may have been worse off than former 'middle' peasants on average, a significant number of those in the 'poor' peasant category were now surplus producers. By 1959 over half the surplus producers were former 'poor' peasants, indicating a high degree of socio-economic mobility to which demographic factors would have been a major contributor. The source data for Table 4.6 are cited by Truong Chinh to show the 'tendencies to change in (peasant) class status.'[47]

Table 4.6: Rural Structure, March 1959

Family Classification[1]	Surplus Families	Adequate Families	Deficient Families
	as % of all rural families		
Landlord	0.4	1.1	0.6
Rich peasant	0.5	0.5	0.2
Middle peasant	8.9	17.8	5.2
Poor peasant	12.0	32.0	16.8
Total	21.8	51.4	22.8

Note: 1. As fixed during Land Reform

Source: Derived from a 'Study of Levels of Nutrition' in thirty-four villages in five provinces, cited by Truong Chinh (1959), pp.9-10 (in which the standard used for 'adequate' is undefined) and from family classification data in Tran Phuong (1968).

This evidence of changes in socio-economic status over the period from 1953 to 1959 sheds light on the problems of classification and land distribution discussed in Chapter III. These problems derived from the assumption of a fairly constant rural stratification. It also suggests that initial landownership was not the only factor contributing to differentiation.

Table 4.7 shows official data on incomes per head.[48] It shows a reduction in differentials between the different strata in the years before collectivisation, based on classifications fixed during Land Reform. Average 'poor' peasant income rose by 18% between 1957 and 1959, while 'middle' peasant income rose by only 5% and 'rich' peasant income declined by 5%. Table 4.6 has already shown the inadequacy (by 1959) of the classifications fixed during Land Reform. The apparent narrowing of differentials may result from socio-economic mobility--'rich' peasants can only go down, 'middle' peasants can go both ways.

The gap between a 'middle' and 'poor' peasant in 1959 may well have been greater than in 1956, but the 'middle' peasant of 1959 may have been a 'poor' peasant in 1956. This does not negate the contention also stressed by Truong Chinh that 'poor' peasants were a motive force behind collectivisation, but indicates that the 'poor' peasant strata of 1959 was not necessarily synonomous with that of 1956 or indeed of 1953 or 1945. Chapter III discussed difficulties consequent upon mobility in these earlier periods.[49]

Land ownership and capital resources may have influenced the continuing process of socio-economic mobility and differentiation but other factors such as differences in ability, education, and experience, as well as family labour resources and age structure were also involved.

<u>Table 4.7: Income Per Head, 1957-61</u>

Family Classi-fication[1]	1957 dong	1957 index	1958 dong	1958 index	1959 dong	1959 index	1960 dong	1960 index	1961 dong	1961 index
Average	129.3	100	126.9	100	143.3	100	130.5	100	137.7	100
Poor peasant	115.6	89	112.5	89	136.7	95	118.2	91	128.9	94
Middle peasant	140.0	108	128.3	101	146.6	102	142.2	109	148.6	108
Rich peasant	187.3	145	169.6	134	178.6	125	174.6	134	181.4	132

<u>Note</u> 1. As fixed during Land Reform.

<u>Source:</u> Nguyen Chi Thanh (1963a), p.232.

Further evidence on differentials apparently came to light with the currency conversion of March 1959. One new _dong_ was issued equivalent to 1,000 old _dong_ and high value notes had to be exchanged immediately, after which only notes of less than 200 old _dong_--sufficient to buy 1/2kg of rice--remained in circulation. The exchange of currency revealed evidence of a concentration of cash in the hands of 'rich' peasants and some former 'landlords.' These assets would have allowed their owners to engage in usury and speculation as well as to purchase land and other property from those in difficulties.[50]

The data in Table 4.8 are cited in support of this argument. Although they show substantial amounts of cash in the hands of 'rich' peasants (officially only 1 percent of the rural population in 1959), they reveal only minimal differentials between 'poor' and 'middle' peasants. The differences would be lower in per capita terms. In addition income and assets are not necessarily correlated. Instances of land, property, and livestock being sold are also cited in support of this argument. However, the data presented show that, while more 'poor' peasants were involved, they were not disproportionately represented. For example, a study in May 1958 of twelve villages in four provinces shows that of 119 families that had sold ricefields, 62% were 'poor' peasants, 24% 'middle', and 1% 'rich' peasants; the remaining 13% were 'small traders.'[51] These accord broadly with the fixed classification proportions. Moreover the number of families involved was a small fraction of the total within the villages. A family with more land than could easily be farmed might, depending on market conditions, choose to sell rather than to rent it or hire labour. Other examples cite instances of buffalo being sold; not only was this often a seasonal practice, but also only upper 'middle' or 'rich' peasants would have owned a buffalo. Given the time periods involved, these examples seem to be further indicators of intra-peasantry mobility.

The possibility of 'speculation' in agricultural produce and the ability of surplus producers to withhold supplies from the State would seem more plausible reasons for concern over accumula-

Table 4.8: Peasant Family Cash Reserves, March 1959 (new *dong*)

Family Classification as fixed during Land Reform	Tien Long district in Kien An province	Binh Luc district in Han Nam province: Rice growing areas	Binh Luc district in Han Nam province: Rice and subsidiary crop growing areas
Poor peasant	55	32.5	66.6
Middle peasant	62	32.5	97.2
Rich peasant	584	42.8	420.9
Landlord	90	....	.....

Source: Truong Chinh (1959), pp.8-9, citing report of National Bank of Vietnam.

tions of cash and capital in rural areas. That differentials existed in the rural sector cannot be denied. What is significant, however, is the extent of this differentiation, the degree to which it was increasing, and the degree to which it was based on post-Land Reform differentials rather than on continuing socio-economic mobility within the peasantry. The available evidence appears to support neither the contention of a pronounced tendency to spontaneous capitalist development[52] nor that individual exploitation would rapidly lead to differentiation.[53]

The leadership was also concerned about the consequences of differentiation for poor peasant support for the VWP. In addition they believed that the State could not both limit the process of differentiation and take advantage of the positive aspects of individual peasant production.

> In reality the power of the State could not impede the objective process of differentiation of classes in a regime of individual exploitation. A new class of rich peasants would appear and accompanying it the regime of landlordism and usury; a class of middle peasants would follow and consequently political instability would be the fundamental characteristic; then one would see the

rebirth of a class of poor or landless peasants, resigned and full of hate against the regime.[54]

The third major rationalisation for cooperatives was related to the overall development strategy, the relationship between agriculture and industry, and the crucial issue of procurement. This argument centered around the alleged contradiction between a 'socialist' industrial sector and a small scale peasant agricultural sector.

The contention was that for a planned industrial sector to develop it could not rely on an unplanned agricultural sector to supply it through the mediation of the market with the necessary foodstuffs, raw materials, and foreign exchange derived from agricultural exports. For the nascent State this was one of the lessons learned from the previous three years and from the problems of prices and procurement discussed earlier. Agricultural development was necessary to improve the living standard of those within agriculture but was also important for industrial development. In the immediate period the main source of accumulation was agriculture. Foreign aid was another source but its impact on agriculture in a static sense depended on whether this was treated as additional to, or a partial substitute for, the taxation of agriculture.

By providing greater direct control over agricultural output and its distribution, collectivisation would supposedly curtail the ability of the peasantry to withhold produce and prevent 'rich' and 'upper-middle' peasants from extensively influencing market prices and quantities. Control of trade was important in this respect, but the preceding years had shown that control of production was also required to suppress the open effects of the market mechanism. Thus, the policy statements of late 1958 and early 1959 placed great emphasis on the complementary roles of the two sectors, the need 'to educate the peasantry with respect to their role,'[55] and the problem of mobilising resources from a multitude of peasant small-holdings.[56]

Experience with limited numbers of cooperatives and MATs during the previous years also influenced the timing of collectivisation. In addition Truong Chinh argued that the VWP had been slow in responding to pressure for collectivisation from sections of the peasantry. The opposite tendency of cadres pushing peasants into cooperative forms of organisation faster than they wished was also noted. This indicates a divergence of views both between cadres at different levels and within the peasantry.

Given these varying pressures, the VWP had to indicate whether collective production was to be supported and directly stimulated or whether family production, backed by MATs, credit cooperatives, and supply and marketing cooperatives was to provide the basis of the agrarian economy. The VWP opted for rapid collectivisation. Its decision was influenced by several factors: first by the difficulties with regard to pricing and procurement; second by the conviction that cooperatives were an inevitable feature of socialist industrialisation; and third by the wave of over-optimism about all forms of socialist and cooperative production heavily inspired by contemporary developments in China.

The adoption of absurdly unrealistic targets was partly caused by cadres at different levels responding to the optimism of the leadership. Results became exaggerated, as did future targets and the emulation campaigns based on them. The aggregate derived from these targets plus an additional margin added at each level produced, for example, the 1960 paddy target of 7.6 million tons. At a conference in September 1959 on crop production at which 'leading officials from each province repeatedly amended the targets for next year's crop yield,' Le Duan said that,

> Before, we thought that we could only increase our yield with tractors, but the Chinese Revolution has set a good example: with determination and diligence, with or without modern tractors, the peasants can still increase the yield to 10 tons of paddy per hectare.[57]

The average yield on two-crop land was some

3.5 tons at this time and it remained below five tons with the exception of 1974. By 1975 only 40% of cooperatives had achieved yields of five tons.[58] Comparable examples can also be found in other sectors of the economy. The target cited by Le Duan was in turn based on false Chinese data.

The evidence for differentiation is open to differing interpretations. The case for cooperatives--as opposed to other forms of agricultural organisation--assisting the development of techniques and raising production and productivity is at best unproven in the DRV or elsewhere. The available evidence about the third rationalisation is clearer. As seen by the State, a problem did exist if it were to proceed with its planned development strategy. The possibilities of adopting alternative approaches to development or of pursuing socialist industrialisation in a different manner based on other forms of agricultural organisation--even if explicitly considered, were overwhelmed by the atmosphere of the period, as well as by the leadership's apparent belief in its own and some of the Chinese rhetoric, in addition to the weight of its ideology.[59]

By 1959 the decision to proceed with rapid collectivisation had been made and it began to be implemented. It was justified on the grounds that it was 'progressive,' that it could both substitute for and facilitate the development of techniques, that the alternative was inevitably rural capitalism and, most convincingly, that only through direct control could sufficient agricultural produce be obtained by the State at prices of its own choosing. However, it was emphasised that the 'principle of voluntarity' was to be followed and that it was necessary to convince the peasantry of the merits of cooperatives rather than to employ coercion. The interests of different groups within the peasantry were thus significant in influencing the development and progress of the movement.

The 'class line' to be followed in the formation of cooperatives was:

> To rely completely on the poor peasants and lower middle peasants, closely unite

> with the middle peasants, restrict and then eliminate economic exploitation by the rich peasants, frustrate the intrigues of the former landlords, continue to open the way for them to work and transform themselves into new men, resolutely guiding the peasants along the path of agricultural cooperation and taking them to socialism.[60]

Certain problems in interpreting this line were inevitable since classification had remained ostensibly unchanged from Land Reform, despite the obvious changes in individual circumstances and the problem of defining upper-middle peasants.

The first cooperatives had been established during the early stages of the liberation war, mainly in the area that became South Vietnam. These experiences had certain influences on later developments, one of which was the recognition of the need to develop agricultural cooperation in a systematic manner. The tendency for cooperatives to be characterised by an over-emphasis on capital utilisation rather than labour utilisation was also noted, even at this early stage.[61]

The first six, post-1954 experimental cooperatives were set up in 1955. By the end of 1956, thirty-seven were in existence and reportedly had paddy yields 14% higher than peasants working individually. With the admission of 'errors' and subsequent Rectification, the movement slowed down as the MATs had done. In October 1957 there were thirty-three cooperatives, although some of them were recent formations. They were small organisations containing a total of only 446 families and 316 hectares.[62] The total number rose steadily during 1958 with major development occurring during the 10th month harvest and coinciding with the final burst of development of MATs. By the end of 1958 there were nearly 5,000 cooperatives representing almost 5% of peasant families with an average membership of twenty-six families. Of the total, twenty-nine were of the high-level variety (in which land was communally owned) with an average membership of some thirty-eight families. The average cultivated area of a cooperative was less than twenty hectares.[63]

By May 1959 when Truong Chinh's speech was delivered to the National Assembly, the number of cooperatives had doubled to over 10,000. Targets for the end of the year had been set at 50% of peasant families in low-level cooperatives and 4% in high-level ones. Some smaller cooperatives had amalgamated, thus raising the average size to approximately forty families.[64] This rapid growth continued after the 16th Plenum of the 2nd VWP CC and the National Assembly meeting. As the number of cooperatives increased, their consolidation required additional attention and was dealth with at some length in the speech by Truong Chinh. His emphasis was on the need for the VWP to control the movement and provide clear guidance. This was also a main thrust of the 16th Plenum and National Assembly resolutions.[65]

With the position of the VWP made clear, the movement progressed rapidly. By the end of May there were over 15,000 cooperatives comprising almost 20% of peasant families; by the end of June, 16,200--21.3% of families. Following the emphasis on consolidation in August and September 1959, the total in late October rose to almost 23,000, comprising over one million families or 38% of the total number of peasant families. There were 565 high-level cooperatives in mid-September 1959.

A further acceleration of collectivisation in the last three months of 1959 was again preceded by discussion which culminated in a seven-day agricultural conference in Hanoi in September addressed by Premier Pham Van Dong, Ho Chi Minh, and Truong Chinh. This further emphasised the consolidation of existing cooperatives and the application of the class line.[66] Consolidation was important given the rapid formation of cooperatives by cadres swept along by the rhetoric and optimism at higher levels. Many of these may have had only a nominal existence initially. Experience with MATs, many of which disbanded during the Rectification period, would also have made the VWP leadership aware of the need to firmly consolidate cooperatives so that a similar process was not repeated.

By the end of 1959 there were 28,775 cooperatives of which 1,033 were high-level. They included over 45% of all peasant families.[67] Significantly, in light of the continuing procurement difficulties, only some 10,000 had been formed in time to play a major role in the cultivation of the bumper autumn rice harvest.[68] The general picture also conceals regional variation where, for instance, the mountain areas in particular were some distance behind the delta.

There was little change in the extent of collectivisation in the first three months of 1960. Consolidation was again emphasised and the formation of new cooperatives received a low priority. The catastrophically low 5th month harvest in 1960--the first to which cooperatives had made a major contribution--had affected the situation in a number of ways. One was the increased attention devoted to food procurement and distribution issues. The low harvest was also a salutary reminder of the short-term reality of agricultural production and potential as well as the organisational problems of cooperatives, especially with respect to labour. It further showed the need to draw the surplus-producing peasants into cooperatives, to benefit from their resources, experience, and ability and, most importantly, to gain access to their output.

Despite problems inherent in an attempt to classify both the rural structure in 1959 and peasants' varying interests and attitudes towards cooperatives, the VWP tried to relate the two together. This influenced the development of cooperatives, especially as the 'class line' presupposed different material interests and endeavored to take account of them. Two-thirds of the positions on the managing boards of cooperatives were to be held by 'poor' and 'lower-middle' peasants with 'poor' peasants predominating. Other 'middle' peasants were to hold the remaining one-third of these positions. Wherever possible, the leading posts were to be held by 'poor' peasants. Family classification remained as had been fixed during Land Reform. These proportions reflected the view that 'poor' and 'lower-middle' peasants constituted some 60% of the rural population, while other 'middle' peasants constituted some 30%. 'Rich' peasants --initially to be excluded from cooperatives-- constituted only 1% and other rural groups the remainder.[69] This would ensure both the recognition of the 'middle' and 'upper-middle' peasant interests--in terms of surplus they were the important groups--and that they did not gain disproportionate influence within cooperatives. 'The administrative ability and production experience of middle peasants--upper-middle peasants in particular--made them natural candidates for leading positions in the cooperatives.'[70]

The use of the term 'middle' peasant to refer to an amorphous sector of the peasantry sometimes distinguished from 'lower-middle' and at other times from 'upper-middle' is indicative of the problems of classifying a peasantry in which only limited differentiation existed, and which was itself in a continual state of flux. The changing composition of these strata is recognised explicitly by Truong Chinh who refers, in the context of the 'class line,' to reliance on 'poor' and 'lower-middle' peasants 'both old and new.'[71] This category thus included the following: those classified as 'poor' and 'landless' peasants during Land Reform, those in the lower part of the 'middle' peasantry classification, and those who had experienced a deterioration in their living conditions since Land Reform and were now effectively 'poor' peasants. The data in Table 4.6 suggest that the latter could have come from any of the rural classes.

Thus there was a very narrow line between ensuring that 'upper-midddle' peasants joined the cooperatives by recognising their interests, and ensuring that they did not take over the cooperatives. The former involved using minimal pressure, although the 'voluntary principle' did not exclude the use of a range of 'incentives' including discriminatory pricing and provision of inputs. The interest rate charged on loans from the State Bank to cooperatives was 6% per annum, while private cultivators paid 10.8% per annum.[72] The 'principle' was itself violated in some areas.[73]

If 'upper-middle' peasants dominated the

cooperatives and used them to protect peasant interests vis-a-vis the State in general and their own interests in particular, then from the State's viewpoint, this would defeat a major purpose of their formation. This was a serious possibility because of the limited VWP rural membership which included cadres with divided loyalties.[74]

The 'official' rural classification in 1959 appears to have been 60% 'poor' peasants, 30% 'middle,' and 1% 'rich' peasants (the others being non-agricultural families). On the basis of this and other data,[75] as well as the definitions of rich, middle, and poor used earlier (see Chapter III), some analysis of likely attitudes towards cooperatives can be attempted.

A substantial section of the peasantry--at least 50% and possibly up to 60%--had landholdings and other resources which made their family living conditions precarious. In good harvest years they might produce yields sufficient for adequate year-round consumption, while in poor years these yields would be insufficient. Their market involvement would be limited to exchange for 'necessities' not produced within the family, assuming roughly constant output. The abolition of landlordism and widespread usury and the re-introduction of tax payment in kind would actually tend to reduce their market involvement, although they might still have to resort to moneylenders to finance their consumption deficit. The price inflation of 1956-58 would have affected them adversely although their only responses would have been delays in tax payment--avoidance if possible --and the expression of discontent within peasant organisations including the VWP.

Within this group there would have been some who had become landless since 1956, although the virtual absence of explicit reference to this suggests such a group was not numerous. There would have been some who had been poor peasants since before Land Reform, and who had yet to see significant gains in material terms from the revolution. The driving force for cooperatives _may_ well have come from this group. However, poor peasants are _less_ likely to be prepared to risk dependence on the labour of others and are likely to be highly suspicious of the dominance of co-operatives by richer peasants. The 23% of the rural population shown in Table 4.6 to have consumption deficiencies would have constituted the lowest sub-grouping within the poor peasantry, while not necessarily being co-determinous with either the landless, those poor since 1953 or earlier, or the activists for collectivisation.

Cooperatives could offer the poor peasantry greater security than individual cultivation by reducing risk from random factors affecting some, but not all families. However, they could not reduce the risk in bad years unless State compensation acted only through the cooperatives. The middle peasants, some 15% to 25% of the rural population, may have shared this interest in co-operatives as an insurance against random calamities.

Different family circumstances and market conditions would affect attitudes of poor and middle peasants. A family lacking labour but with a small amount of land would lose if the rental payment from the low-level cooperative was less than that from private tenants or the return from the use of hired labour. In a high-level cooperative making no rental payments, the family would be worse off unless the cooperative achieved significant output and income increases. The same family might also be worse off if less flexible cooperative working hours adversely affected family sector output or family income.

Cooperative membership might be attractive, however, especially to those with above average land, in that it 'guaranteed' landownership and hence rental income in a situation in which assets and income could fluctuate considerably over a short period. This rental income was one possible source of conflict between peasants.[76]

A family lacking land relative to labour resources would lose if the return from working for the cooperative were less than that for working elsewhere. Again, fixed working hours might reduce the family's ability to maximise the total return rather than the average return to labour time. A family with above average land and a favourable labour/consumption balance might be constrained by input availability and would join a cooperative if this provided preferential access to inputs, including credit.

These examples serve to illustrate that attitudes towards collectivisation depended on family endowment which included land, labour, and other resources, the age structure of the family, as well as its current consumption level. Peasant attitudes would also be affected by a number of other factors such as the aspiration for land ownership; the desire to maintain individual control over resource allocation and income distribution; the opportunities for mobility in which the possibility of upward mobility may have weighed more than the fear of further deterioration in living standards; and the fear of free-loaders benefiting from the labour of others. The official disparaging view of the peasantry summarized these factors as follows: 'inside each peasant the ideology of a small owner remained deeply rooted for a long time.'[77] The 'ideology' derived from material interests and it remains deeply rooted in Vietnam. The degree to which the alternative ideology (which argued the greater productive potential of cooperatives) was accepted, could also influence peasant attitudes.

The rich peasants, who probably comprised between 10% and 15% of the peasantry, possessed sufficient resources to maintain their position as surplus producers, though the degree of mobility into and out of the strata may still have been high. It would have included some former 'rich' peasants, 'landlords' who retained some land and capital, the more favourably placed members of the 'middle' peasantry, and even some who had been classified as 'poor' peasants during Land Reform. The good harvests after Land Reform could have resulted in families which became relatively rich in terms of income but remained 'middle' in terms of the official classification which was concerned primarily with land endowment. They would have been able to lend money and trade in their own and others' produce. Those who had increased their holdings of land would have been employing additional labour. Control over this group, if not their support, was essential in the eyes of the

VWP leadership.

Since the security question was less important to them, this group would have opposed collectivisation in principle. Their opportunities for moneylending and hiring of labour would be reduced (their employees might well join cooperatives), as would their trading activities. Their larger per capita landholding would not receive proportionately as much in terms of rental income, and their role as organisers and administrators might be limited if they were viewed with mistrust. However, many of the poor and middle peasant attitudes also apply in the case of rich peasants.

From the viewpoint of the VWP, the risk of having this group dominate the cooperatives was larger given their greater resources--they had more implements, they tended to have more family members, and often more ability. Hence the policy line was to 'eliminate economic exploitation by the rich peasants' and, initially, to exclude them from the cooperatives.

The classification of the peasantry into three groups in 1959 is an extreme oversimplification. The groups given here are based on overall endowment. They differ from those in the official classification which was based primarily on land and method of 'exploitation,' by identifying a rich peasantry significantly larger than the 1% or 2% 'rich' peasants during and after Land Reform.[78] An alternative stratification (derived directly from Table 4.6) would show a plausible distribution made up of poor and rich strata, each consisting of 25% of the peasantry, and a large middle strata with the remaining 50%. One DRV estimate shows a 'middle' strata of over 60% in 1959.[79] The decision not to attempt a new classification[80] confirms the inadequacy of the old one. It also points out the difficulties inherent in classification, particularly if the 1953 criteria and their limited usefulness were retained.

Drawing a correspondence between any of these stratifications and attitudes towards cooperatives is impossible. Attitudes varied according to constraints which varied within the three groups and their sub-groups. These variations related to the composition of the family endowment rather than to its size, as well as to other factors. The effective classification actually employed by the VWP was relatively straightforward: all who were less than enthusiastic about cooperatives, as conceived by the VWP leadership, were 'upper-middle' or even 'middle' peasants.

The consolidation of cooperatives was more significant than their formation for a number of reasons. First, the nominal existence of a cooperative did not guarantee that it would function as a cooperative, in terms of labour organisation in particular. Of some 25,000 cooperatives in October 1959 only 10,000 had functioned during the bumper 10th month harvest, although 15,000 had allegedly existed since the 5th month harvest.

> In the labour management field, most cooperatives were still at the phase of common discussion and evaluation of work. In general, both production and labour management in newly set up cooperatives was confusedly carried out and remained at a rudimentary phase.[81]

Second, experience with MATs during 1956 and 1957 suggested that the failure to consolidate had resulted in rapid disbandment. From a peak in mid-1956 of over 150,000, the number of MATs had slumped to less than half this number twelve months later. Third, only consolidation could ensure that the cooperatives fulfilled their function as defined by the VWP. Violation of the 'class line' could result either in an over-attention to 'middle' peasant interests[82] or in the reverse--an alienation of the most important per-family surplus producers.

One problem for the leadership was that rural VWP cadres also had varying attitudes towards cooperatives. They were drawn from different stratas and also had strong ties to their families in terms of aspirations and attitudes. Their viewpoint was thus a major determinant in the pace of collectivisation in different regions and villages. It also affected the fine balancing of interests required, and hence the distribution of authority within any particular cooperative. The role and influence of middle peasants within the VWP had already led to the admission of 'errors' in Land Reform and Rectification. This had consolidated their position politically, while the removal of landlords and the effective neutralisation of rich peasants had raised their relative economic status.

Consequently, any overall picture of the DRV rural sector during the development and consolidation of cooperatives neglects substantial local differences which the preceding education and ideological campaigns had not eliminated. Thus, in some areas VWP branches pursued collectivisation vigorously, applying coercion to peasants to ensure they joined and violating certain peasant interests within the cooperatives by, for example, minimising rental payments. In other areas VWP branches may either have opposed collectivisation altogether, at least prior to 1959, or ensured that cooperatives expressed particular interests above all others. In the official interpretation these practices were associated with 'poor' and 'middle' peasants respectively.[83]

The VWP leadership decisions and detailed rationalisation of immediate collectivisation in late 1958 and early 1959 ensured that the movement continued and that existing progress was consolidated. It also endeavoured to maintain a balance between the varying and conflicting interests.[84] VWP leadership control over collectivisation ensured that it continued, since the leaders believed that the building of socialism demanded as an inherent characteristic, a collectivised peasant agriculture. This control also ensured that it did not turn into a more extreme class conflict and alienate the middle peasants as had occurred during Land Reform. Peasant support for cooperatives in terms of material interests and aspirations was less than it had been for Land Reform, and the consequences of an extreme policy could have been even greater than in the earlier period. The absence of active opposition to collectivisation may also have indicated a widespread preparedness to at least try the new system. These were years of peace, good harvests, and expanding industry after years of war and instability. There was also a momentum for agrarian change.

By 1960 the extreme optimism of the earlier period had subsided for a number of reasons. One of these was an increasing awareness of events in China which had a direct consequence for the DRV in terms of aid. Experience with cooperatives for at least one paddy harvest (at 1.16 million tons, a disastrously low one) in many areas, and more harvests in others, had revealed a range of problems which, although partially recognised or foreseen earlier, had a cumulative impact.

The continuing difficulties with procurement increased the need to include middle and rich peasants in cooperatives, but also drew attention to their ability to dominate the cooperatives and use them to protect peasant interests. These peasants had to be attracted into cooperatives to gain control of their resources, ability, and output. However, even with the extensive use of economic and other incentives, the possibility was limited for demonstrating 'the obvious superiority raising the cooperatives' output and their members' income.'[85] The danger of cooperatives asserting their autonomy was shown by the withholding of produce from the State in order to sell it at higher prices on the free market.[86] This action reflected the peasantry's conception of their material interests. The State believed the peasantry should make sacrifices in order to build socialist industry which would be in their long-term material interests.[87] The rise in the free market rice price affected peasants in deficit regions and, more significantly, the urban workforce. This increased the pressure by the State for control of commerce and production.

Emphasis during this period was placed on increasing actual cooperative production. Pham Van Dong, in April 1960, referred to members of cooperatives who worked only one hundred days per year.[88] Ideological education of cooperators was also stressed--to induce them to work collectively and to fulfill State demands in the interests of the whole country (as perceived by the State). Consolidation involved neither a reconsideration of collectivisation as a means of fulfilling objectives nor explicit consideration of alternatives. The construction of the rationale for collectivisation and its interpretation that a dichotomy existed between two paths--'capitalism or socialism'--effectively precluded the possibility that there might be more than one means of building socialism.

In the final months of 1960, the focus was on incorporating the majority of the peasantry into low-level cooperatives, particularly the middle and rich peasants. These groups had not yet joined in significant numbers, despite claims that they dominated cooperatives in certain places.

Reference to 'middle' peasant dominance of cooperatives ascribes all expression of independent peasant economy interests to this group. It was difficult for the VWP to recognise a conflict of interests between the peasantry and the State, which was ostensibly based on an alliance of workers and peasants. A more explicit recognition of such a conflict can be seen in the following comment from 1964.

> We must acknowledge that 50% of the Party members in the countryside (including the average and poor Party members) have not definitely resolved the question of 'who emerges victorious' with respect to ideological matters touching on their own selves.[89]

It was reported in March 1960 that:

> The majority of persons who do not belong to cooperatives are middle and upper-middle peasants. The reason these people have not made up their minds to join cooperatives and have manifested a 'wait and see' attitude is the fact that they possess more production means than poor and lower-middle peasants. Therefore, it is necessary patiently to persuade them of the cooperatives' advantages.[90]

Official data show that 'middle' peasant income from cultivation was 10% lower in the cooperatives than outside. Consumption levels of cooperators were also below those of independent 'middle' peasant producers.[91] These data, the above report, and the data in Table 4.7 suggest a widening of differentials between those outside and inside the cooperatives. This would make the task of 'persuasion' extremely difficult. At the end of 1960 with 86% of the peasantry in cooperatives, the 14% of families outside of cooperatives held 32% of the cultivated land, suggesting that those rich in terms of land alone remained unconvinced of the merits of cooperatives.

The major drive to complete collectivisation coincided with the 10th month harvest of 1960. Although not as high as those of 1958 and 1959, this harvest was still high by historical standards (see Table 4.4).[92] It was accompanied by moves to raise the average size of cooperatives--to benefit from increasing returns to scale[93]--and to increase the number of high-level cooperatives.[94]

By the end of September 1960 over 75% of peasant families were in cooperatives--over half a million having joined in that month alone.[95] The proportion rose to 81% in October and at the end of 1960 there were more than 40,000 cooperatives containing almost 86% of peasant families and 68% of the cultivated land. The average membership was fifty-nine families, although the high-level cooperatives which contained 12% of peasant families[96] were usually larger. Significant regional variations in the size of cooperatives and the rate of their formation persisted.

The preoccupation with consolidation to increase the extent of collective operations, improve organisation, raise the technical level, and ensure that cooperatives fulfilled State commitments increased with the number of cooperatives. Transformation from low- to high-level cooperatives emerged as a major issue after 1960.

The fall in production in 1960 was considered a setback resulting from climatic factors and inadequate organisation of cooperatives which could be overcome. More attention was paid to organisation in subsequent years. In retrospect, the year 1960 marked a shift in the trend of agricultural production, of paddy in particular. Whereas paddy output had risen by almost 50% between 1955 and 1959 to reach a level of 5.2 million tons, it fluctuated around a level of 4.3 million tons for the next fifteen years.

The assertion of Truong Chinh that

> the small peasant economy in general is not only unable to extend production every year, but is unable to achieve simple reproduction in every crop... small peasant economy is scattered and precarious, usually heavily influenced by the market and much dependent upon natural conditions[97]

is difficult to reconcile with experience during these years. The growth rate of the late 1950s was probably not sustainable, since part of it represented a recovery from the years of war. Subsequent performance was, however, poor and had two contributory causes: the changes in the overall economic situation which constrained agriculture and made growth more difficult, and the problems caused by collectivisation.

In accordance with the third argument for collectivisation, the autonomy of the peasantry to control the marketing of their produce was severely curtailed in 1960. However, the taxation of agriculture remained a major problem in subsequent years, partly because of difficulty in establishing the first argument.

Summary

Chapter IV began with the period of rising production after Land Reform and showed that this produced the DRV's own scissors crisis.[98] The procurement problems of these years were a direct stimulus to the State's desire for collectivisation.

As in China, the DRV went from Mutual Aid Teams (MATs) to low-level and then to high-level cooperatives. The first stage--MATs--was shown to have been delayed by the Rectification Campaign. The break-up of many MATs during this period is used as evidence of peasant antipathy towards forms of collectivisation.

The leadership rationale for collectivisation was dominated by three arguments: that cooperatives were more productive than private agriculture; that they would prevent differentiation, leading to capitalist agriculture; and, that they would secure central State control over agricultural production and surplus.[99]

Chapter IV shows that there was, and still is, no evidence for the validity of the first argument, particularly in the absence of mechanisation. The empirical evidence cited by the leadership in support of argument two--that differentiation derived from the post-Land Reform landholdings was increasing and would lead inexorably to capitalism--is shown to be compatible with different interpretations. While the evidence is limited, it must be assumed that the leadership selected that which they believed most clearly substantiated their argument. Therefore, it appears unlikely that 'new' evidence would counter the conclusion that the leadership's argument is at best unproven.

Argument three was derived directly from procurement problems and the need to prevent agriculture from acting as a constraint on industry. The theoretical background for this was examined in the Introduction. The extent to which the State succeeded in controlling agricultural production and surplus is examined in subsequent chapters.

The formation of low-level cooperatives was marked by bursts of creation interspersed with consolidation, as was also the case in China.[100] The extent to which cooperatives were a reality was raised during this chapter and will be considered in subsequent chapters. It was emphasised that the existence of a cooperative in the records did not guarantee that it functioned either as a cooperative or in the State's interest.

A further attempt was made to consider peasant classification and peasant attitudes towards collectivisation. It was argued that these were affected by a range of factors, not all of which are empirically observable--endowment, age and sex composition, aspirations, local political and geographical situation, ideology. The 'official' attitude was that those in favour of collectivisation were 'poor' peasants and those ambivalent or against it 'upper-middle' peasants. It was argued that this dichotomous view had no basis in reality.

It has also been shown that the State and VWP were aware of the need to maintain a balance between over-zealous, rapid formation which could alienate at least sections of the peasantry, and too slow formation which would not serve the State's objective of control. The State also had to maintain a balance between incorporating surplus producers into the cooperatives and ensuring that they did not take over the cooperatives. Depending on the VWP strength and on cadre attitudes, instances of violations of these balances occurred. As in the case of China, they were construed as 'leftist' and 'rightist' deviations and it was probably better to err on the 'leftist' side.[101]

This chapter concluded with a consideration of the problems emerging in 1960. The limited ability of cooperatives to function effectively and expand production reduced the incentive for peasants to join and participate enthusiastically. This made it all the more essential to incorporate surplus producers in order to obtain access to their resources, ability, and production. However, it also meant that they could not be alienated or production could fall. The incompatibility of these varying objectives, the divergence of peasant and State interests, and the repetition of the collectivisation arguments are pursued in subsequent chapters.

CHAPTER V

Emerging Problems and Policy Responses, 1961-65

Introduction

Chapter IV reviewed the arguments adduced in support of collectivisation, the formal establishment of cooperatives, and the recognition of peasant interests that this involved. It emphasised that the mere formation of a cooperative did not ensure that it functioned as a cooperative in labour organisation, management, production, or distribution. The attempt to make cooperatives function as collective organisations and fulfill the role laid down for them by the State was one feature of the period 1961-65.

A second major feature which was related to consolidation was the attempt to demonstrate the alleged superiority of cooperatives. This task would be made easier if cooperatives could either stimulate or replace the need for new techniques and increase production. An obvious problem was that antipathy on the part of the peasantry and reluctant participation in cooperatives made it more difficult to 'prove' their superiority. This strengthened the case for rapid consolidation.

The third dominant feature of this period was concern over the mobilisation of agricultural produce which had been a significant factor in the formation of cooperatives. Consolidation and higher production alone would not ensure that cooperatives fulfilled the interests of the State as opposed to those of their members in those cases where these interests diverged. Thus control was necessary, coupled with attempts to convince the peasantry that their interests--at least in the long run--were being served by the State and VWP. Where control was inadequate (for VWP penetration of rural areas was limited and the attitude of many cadres ambivalent), pricing policy and the provision of incentive goods would play a part.

These three themes--consolidation, collective superiority, and procurement--are developments of the three arguments for cooperatives. Consolidation would counter the tendencies toward differentiation and develop collective, as opposed to individual, production. As in previous years, the resulting policies also had to be cognizant of the different interests within the peasantry. Thus, consolidation from low- to high-level cooperatives had to take into account those peasants who would lose rental income. More crucially, the need to maintain food mobilisation, especially in the face of stagnating production, required recognition of the importance of individual production, both inside and outside the cooperative. Failure to bring about increases in agricultural production through collectivisation, at least in line with population increases, would make procurement more difficult, and put pressure on the already low living standards in rural and urban areas. Rapid industrial growth or unlimited foreign aid would have eased this situation by relaxing the constraint on the availability of industrial goods and limiting the squeeze on agriculture.

Thus, policies had to ensure that cooperatives demonstrated their 'superiority,' encouraged enthusiastic peasant participation, and increased both production and the State's share. If this combination could not be found, or, more plausibly, did not exist, then the different goals had to be ranked. The outcome almost inevitably involved a compromise--albeit an unsatisfactory one-- between the State's interests and those of the peasantry, either as a whole or sections thereof.

Cooperatives, Brigades, and Families

Agriculture contained three or four levels of organisation: the cooperative itself, the production brigade, and--in the larger cooperatives--the production team, and the constituent families. This section presents a stylisation of these levels and their interrelationships. The precise details of organisation changed during the period as did the character of the relationships.[1]

Two forms of cooperatives were developed in the DRV. In the low-level cooperatives members received part of the output as a form of rental payment proportional to the amount of land contributed. This dividend was set initially at 25-30% of output in order to attract those with large per capita landholdings by offering competitive rents.[2] In a densely populated area of peasant farming, land rental will always be relatively high (i.e. in excess of its marginal product). In high-level cooperatives, defined as a 'socialist collective organisation,' land payment was eliminated and each member was paid solely according to the labour contributed.

The size of cooperatives had varied considerably both through time and regionally. An objective of the early 1960s was to make them co-determinous with the hamlet (thôn) and later the village (xã). In the larger cooperatives the production brigade was often based on the hamlet. The average DRV village had a population of around 5,000 and was composed of several hamlets. In all there were approximately 12,000 hamlets.

Membership was open to all 'working peasants and other labouring people in the countryside from sixteen years of age upwards' although former 'landlords' had to serve a probationary period. Children of cooperators were automatically eligible for membership at sixteen years of age although they were not obliged to join. Upon joining, cooperators 'must pay fees, pool in their land, cattle, and main farm implements.' Other livestock and tools were not collectivised. Initial moves to collectivise pigs were rapidly reversed.

There were two and sometimes three levels of organisation within the cooperatives. The cooperative owned the means of production. Its executive organ was the managing board which was responsible for the management and utilisation of the cooperative's resources. This body was responsible for relations and contracts with higher levels. It allocated land and assigned tasks to the production brigades.

The production brigade consisted of between thirty and sixty workers in the mid-1960s. Labour was organised at this level with the brigade leadership allocating tasks to individuals or production teams in the case of larger cooperatives. The brigade was responsible for allotting and recording work points and for distributing income in cash and kind. The brigade was also responsible for allocating workdays (unpaid) to public works such as water conservancy projects.

Work was assigned to brigades on the basis of the 'triple contract:'

1. Contract on production norms. Land was divided according to soil and other conditions and norms fixed for each portion, crop, and cultivation method.

2. Contract on the number of workdays and points, based on the quality and quantity of land and work allotted. It was to change when these conditions changed, and was to ensure that the number of workdays assigned to each brigade was proportionate to its membership.

3. Contract on production costs. This was intended to avoid waste in the use of cooperative resources and stipulated the non-labour input to production.

These 'contracts' were clearly open to considerable bargaining between the levels. It was prohibited for the cooperative or brigade to allot tasks to families on the basis of the 'triple contract.'

The current focus of analysis is normally the brigade because the direct labour contact was at this level. Family labour allocation was between family production and the brigade. Weak cooperative management was a problem, from the State's perspective, since this level had no <u>direct</u> control over production and yet it was responsible for fulfilling State obligations. High-level and larger cooperatives with greater VWP and State participation were developed partly to reduce the independence of brigades and their leadership.

Work points were allotted to tasks or types of work, with ten points awarded to the average norm of work--work done by a worker in a day with average labour outlay and technique. The brigade leadership was responsible for recording work days and work points. This presented considerable problems of supervision and was complex and open to 'abuse.' Supervision using 'self-surveillance' was easier in small-production brigades than in larger ones, or in those organised for specific tasks (irrigation or implement repair) where the direct interest of the cooperators in the task was less.

Cooperative members were expected to contribute a specified number of work days per year depending on sex and age. Non-members could also be employed by the cooperative and points earned by them were marked to members for payment. Thus all adult members of a family might not be members of a cooperative while all worked for it at different times of the year.

The main distribution in the cooperative occurred annually, although provisional distributions took place after harvests. All cooperative annual income after deduction of production costs (including seeds and funds for the next crops), taxes to the State, cooperative accumulation (5% to 10%), and welfare funds (2% to 5%), was to be 'entirely shared out to the cooperators in proportion to their work days and work points.' The allocation of output was made by the managing board and distribution by the production brigade. Distribution also attempted to take account of the old, the sick, and families with a low labour/consumption balance. In principle, those who could work should; those who couldn't were covered by social welfare arrangements.

The cooperative was dependent on brigade-level production, and thus targets and performance were potential sources of conflict. Pressures to under-report in order to reduce tax liability and achieve more easily realisable targets would tend to dominate any tendencies to over-report at both levels--brigade to cooperative and cooperative to State. Market conditions influenced these relationships as well as the family/brigade relationship, since the brigade or cooperative might divert production to produce a higher income. In addition to possible conflicting interests between levels, there was the possibility of individual officials using their position to enrich themselves, their families, and others.

Although the VWP leadership had rejected individual family production as the dominant form within the rural sector, it accepted that it had a limited role to play in terms of political expediency and productive efficiency: 'superiority' of cooperatives did not extend to all aspects of production. Within the cooperative, whether high- or low-level, up to 5% of the land was allocated for use by individual families and could be used for the cultivation of non-rice foodstuffs or for livestock. In certain instances more than 5% of the land could be allocated and, in practice, more land had frequently been used for this purpose.

Official policy throughout this period stressed the complementary relationship between the two sectors.[3] While the brigade concentrated mainly on the cultivation of rice, other basic foodstuffs, or industrial crops, the family plot could ensure a supply of other commodities such as vegetables, fruit, and meat, in which cases scale advantages did not apply. In addition, the family plot could make use of the labour time that was not devoted to the brigade. Peasant cash income from their own production constituted a source of demand for consumer goods; some 70-80% of cooperative income allocations were in foodstuffs, mainly rice. The peasantry's attachment to land was also recognised as a justification for the 'family economy'--at least in the immediate future.

> In principle, the land belonging to the members should be entirely transferred to cooperatives for exploitation on a unified basis. However, considering the peasants particular activities and with a view to enabling them to use their labour during their spare time, a plot of land shall be allotted to each family, not exceeding 5% of the per capita cultivable land in the village, to be used for growing various crops (vegetables, fruit trees) and small-scale animal breeding (poultry, pigs).[4]

For the peasant family, allocation of labour time was dependent on its real return in terms of consumption. It involved consideration of the output mix of the brigade and family, the opportunity cost of labour at given output prices, and real market conditions, including the availability of demand for agricultural produce as well as the supply of industrial goods and the relative prices. In practice, family labour was often much more attractive because labour time was more profitable. In addition, cooperative output had a low value and cooperatives were dogged by organisational difficulties, inexperience, and input scarcities.

Thus, there was often a conflict over the allocation of labour between the two sectors rather than the complementarity implied by the official argument. Exhortations were made frequently to raise the number of work days per cooperator. In 1960 it was estimated that only 103 days on average were devoted to collective production. The target was to raise this to 150 and subsequently to 200, and to increase the number of hours worked per day.[5]

The devotion of labour time to the family plot could, under certain conditions, act as a constraint on the cooperative by restricting the amount of labour time available to it. The relatively high proportion of total peasant income derived from the family plot--over 50%[6]--partly reflected the allocation of labour between the two sectors. It also reflected the different valuation of products produced by the two sectors, their different productivities, and the widespread deviation from the 5% limit on land devoted to family cultivation. The devotion of large amounts of labour time, and even land, to famiy production reduced the potential for utilising improved techniques and methods of organisation on a large scale. This limited the ability of the cooperative to demonstrate its 'superiority' in terms of production and productivity, and reinforced the tendency to devote labour to family production.

The family sector could, however, act as a complement to the cooperative sector. By producing substantial quantities of wage goods and selling them to the towns, it could reduce State procurement from cooperatives and hence allow retention of a higher percentage of the surplus. Cooperative production in practice was unable to replace much of the production of the family sector, especially meat.

Secondly, if the marginal productivity of labour time declined as the labour input increased because of the low technical level of agriculture and scarcity of complementary inputs, then the absorption of labour by the family sector would make the value of the brigade work point higher than it would otherwise have been. For the same reason, the brigade had no incentive to take on additional workers. The rise in the average number of work days from 103 in 1960 to 124 in 1961, accompanied by a fall in their average value from 0.80 *dong* to 0.77 *dong*,[7] would support this hypothesis of a declining marginal product of labour time.

> The increase in the number of man days worked per hectare seems to have overtaken the increase in yield per hectare--so that there appears to have been an appreciable reduction in the productivity of the day worked, which has been observed in southern China as well.[8]

The search for an 'optimal' solution to the relationship between the family and cooperative sectors was another theme of this period, and it has remained significant in agrarian policy in the post-1975 Socialist Republic of Vietnam (SRV). The continued existence of the family plot was an economic necessity in the direct sense that it provided a significant proportion of agricultural produce, and indirectly in that attempts to remove it would have produced a peasant response with serious economic and political consequences.

The Early Years

Paddy production in 1960 was a million tons below that of 1959. The State share fell by less than this because of the greater control of trade and the control achieved through collectivisation. Given the rapid rate of population growth and low consumption levels, any reduction in production and procurement was serious. Exhortations to minimise 'wasteful' consumption were repeatedly made. A *Nhan Dan* article of 11 November 1960 referred to the use of 70,000 tons of paddy for the manufacture of alcohol, 100,000 tons for cakes, and 130,000 tons for animal food--the latter representing 3% of human needs. It called for the cultivation of maize, potatoes, and manioc to replace rice in cakes, noodles, and animal foods.[9] The frequent repetition of these calls indicates the degree of opposition to changes in consumption patterns.

Of particular concern was the supply of meat products. As Table 5.1 shows, collectivisation was accompanied by a reduction in livestock numbers compared with the rapid increase during the previous period. The number of pigs fell by almost 250,000 between 1958 and 1961; the number of oxen by over 200,000. Although the number of pigs subsequently increased, the number of oxen remained well below the 1958 level for the entire period to 1975. The fall in livestock numbers also affected fertiliser production since livestock provided the major source of manure. The failure of fertiliser availability to expand along with the increase in cultivated areas during the early 1960s caused fertiliser utilisation per hectare--and hence yields--to fall.[10]

Given the poor quality and intensive utilisation of the soils, fertiliser was essential to maintain land productivity.

These problems had two causes. First, the fall in grain production produced a rise in meat consumption. The second cause was the effect of collectivisation on livestock breeding. Although the official guidelines called for the collectivisation of draught animals only, initially efforts were made to collectivise <u>all</u> livestock. It was soon recognised that animal husbandry was best left to individual families due to the technical difficulties of large-scale animal husbandry with limited supplies of foodstuffs and untrained cadres.[11]

In 1961, 97% of pigs came from peasants' private livestock breeding, the proceeds of which provided 15% of their annual income. The Five-Year Plan target for raising the pig population from 3.7 million to 8 million was based mainly on

Table 5.1: Index of Livestock Numbers, 1956-64 (1958 = 100)

Year	Water Buffalo	Oxen	Pigs	State Procurement of meat (000 tons)
1955	78.8	76.2	53.6	....
1956	84.7	84.1	62.7	....
1957	90.0	91.3	74.0	6.4
1958	100.0	100.0	100.0	30.5
1959	105.2	95.8	91.0	70.8
1960	105.5	88.6	94.1	96.0
1961	104.5	78.7	93.9	56.5
1962	107.6	79.2	106.4	59.9
1963	109.6	80.2	105.6	56.0
1964	113.7	82.6	113.6	73.3

Source: Derived from figures in Vo Nhan Tri (1967), pp. 218, 245, 295, 337, 429, 451. See also Appendix Table V which uses data from other sources.

private breeding.[12] The actual total achieved in 1965 was 4.8 million.[13] Peasant opposition to the attempted collectivisation of livestock may have resulted in increased slaughterings during this period. The higher procurement of meat by the State in 1959 and 1960 (Table 5.1) was due to increased slaughterings, rising meat consumption, greater State involvement in trade, or price effects.

The fall in agricultural production in 1960 affected livestock both directly, by reducing the availability of crops grown purely for feed, and indirectly--particularly in the case of pigs--by reducing the surplus of human foodstuffs available.[14]

The chastening experience of the first year of large-scale collective farming contributed to the assessment of agriculture undertaken at the 5th Plenum of the 3rd VWP CC in July 1961.[15] Increased production was not only a prerequisite for increased procurement, but also for convincing those who had not yet joined cooperatives. The fall in output, the livestock problems, and organisational difficulties had led some peasants to withdraw from cooperatives. Disenchantment with cooperatives and the expression of opposition in some areas concerned the VWP leaderhsip.[16] It was claimed that the percentage of families applying to leave cooperatives never exceeded 1.6%,[17] although this represented a significant number in absolute terms and does not take into account those who never joined (all members of a family did not necessarily join).

The 5th Plenum resolution referred to 'three great contradictions' which had to be solved to overcome the weaknesses of cooperatives.[18] The first contradiction was between the collective form and the technical state of agriculture. The second was the rapid development of cooperatives in contrast to the capacities for technical, organisational, and VWP leadership. There were insufficient VWP cadres to have a cell within each of the 9,000 hamlet-size cooperatives that were being formed in some areas.[19] These contained 150 to 200 families, whereas the overall average, even in high-level cooperatives, was barely one hundred families. The third 'contradiction' concerned the rapid development of the collective form compared with the low level of 'socialist consciousness' of the peasantry.[20]

These 'contradictions' relate to the 'three revolutions'--'transformation of the relations of production, technical revolution, and ideological education'[21]--which was stressed frequently by the VWP leadership, and to the three themes stressed above. The 'contradiction' between the collective form and the state of production techniques could be solved only by the technical development of agriculture. This was also the basis for the alleged superiority of cooperatives. The development of organisational and leadership capacities was necessary to make collective production a reality. It would also consolidate and subsequently transform cooperatives to the high-level variety, as well as ensure that agriculture fulfilled the role specified by the State. That role was to supply food, raw materials, export commodities, and labour, and to constitute a market for industrial goods.[22]

The conversion to high-level cooperatives was, in principle, to occur only when output had risen sufficiently to maintain incomes without the payment of the land dividend.[23] This would ensure that the material interests of those with above average landownership were not violated. In principle, the land dividend was initially 25-30% of cooperative output.[24] Assuming that the larger, owned landholdings were 50% larger than the average, then the overall value of output would have to rise by 15% to maintain income levels without the land dividend. The value of gross agricultural output (at constant prices) did not exceed 115% of the 1959 level until 1972,[25] suggesting that this principle was not always followed in practice.

The third 'contradiction'--the lack of a 'socialist consciousness' on the party of the peasantry--required ideological education. It was related to the objective of incorporating all peasants into the cooperatives and ensuring that collective work became a reality. The inculcation of a 'socialist consciousness' would ensure that peasants both abandoned the desire to work independently and accepted the need to make sacrifices in the interests of the industrialisation of the economy. From the viewpoint of the State, success in this 'ideological revolution' would reduce the need for providing material incentives to ensure that the peasantry fulfilled their prescribed tasks.

The contradiction that spanned all three of these was the one between the collective form and the reality of low agricultural productivity, underdeveloped techniques, widespread family sector production, and procurement difficulties deriving essentially from the limited marketable surplus. This problem dominated agriculture throughout this period.

In 1961 even the first of the 'three revolutions'--involving production relationships--was nowhere near completed. New organisational forms had been established but, unless a purely verbal interpretation is placed on the notion of transformation of the relations of production, a substantial amount remained to be done to make the new form a reality.

The VWP leadership tended to refer to the 'completion of socialist transformation' in the context of the formation of low-level cooperatives,[26] and to talk of a shift of emphasis to technical development ('transformation of the forces of production') as if the two were separate. This indicates either that the leadership was out of touch with reality in the rural areas and believed, on the basis of cadre reports, that 'the system of individual ownership of the small producers...has been turned into the system of socialist collective ownership,'[27] or that they viewed such transformations mechanistically. The latter interpretation is consistent with the argument for collectivisation based on the belief that the cooperative represented a higher level of socialist organisation, the advantages of which would rapidly become apparent and actually facilitate the technical development. Unfortunately, a simplistic interpretation of the relationship between 'relations and forces of production' and the association of focus on the former as 'leftist' and on the latter as 'rightist' has characterised discussion both inside and outside the DRV.[28]

In reality the development of social relations and techniques of production continued to reflect objective circumstances. These included high population density, absolute land scarcity, low agricultural productivity, limited surplus, the material interests of the peasantry, and the interests of the State. Moreover, there were differences within the VWP leadership which were reflected in the focus and emphasis of ministries and individuals. Agricultural development was not confined to 'socialist relations of production,' although it was arguably constrained by the existence of the collective form.

A communique of May 1961 on purchasing policy stated that cooperatives could use production in excess of the State quota for livestock, reserves, or sale on the market. The State would pay ten to twenty percent more for above-quota sales, although this was still below the free market price. At the same time the prices of a number of consumer goods, including books and children's goods, were reduced by between five and thirty percent, in a further attempt to raise the marketed proportion of agricultural production.[29]

The higher purchase prices related to rice, manioc, potato, and cotton. Similar incentive prices for other crops were introduced in early 1962. Reductions in prices of consumer goods could not, however, overcome the problem of their limited supply and, by raising demand, could even worsen the situation. Moreover, the peasantry may still have considered the terms of trade adverse and have preferred to use any surplus to increase their own consumption.

The 10th month crop of 1961 was lower than in the three previous years--see Table 5.2. In early 1962, the State aim was to obtain more from this crop and to ensure a satisfactory 5th month crop in 1962. This crop had already suffered unfavourable weather conditions. The leadership stressed ideological education to encourage greater efforts in cultivation and 'instill a sense of duty towards the State.' It also emphasised crop diversification. State concern over procurement was reflected in the more direct mobilisation methods of even a subsidiary crop:

> If necessary, shock groups of youths, cadres, militiamen, self-defence men and school children, may be mobilised to help gather and process manioc.

In addition incentive purchasing policies were used.

> We must try to buy at an encouraging price the surplus paddy remaining after the fixed amount of food to be sold as a duty has been sold.[30]

Although technical advance was repeatedly referred to as the 'key' to agricultural progress at this time, considerable emphasis was placed on the 'consolidation' of cooperatives.

> A shortcoming has developed which consists in neglecting to consolidate, de-

> velop, and perfect the new production relations. This is reflected in a failure to observe strictly the three principles concerning cooperative management. Cooperative distribution is still characterised by egalitarianism. Ideological and political education has not yet received special attention. Cooperative members still have a low sense of being owners of cooperatives and a low sense of the need to build cooperatives economically.[31]

Egalitarianism in distribution was presumably easier to administer. The system was extremely complex given the shortage of statistical and accounting cadres. At the brigade level it involved overseeing the brigades, making dividend payments for land and animals, deductions for various funds, payment to the State, and then calculating income shares which the brigade distributed. The injunction against egalitarianism recognised its disincentive effects and the danger of antagonising above-average landowners, whether in or out of cooperatives. It also recognised the disincentive effect of equal distribution.

In late 1961 a different income distribution problem emerged. Students and army units were mobilised to assist with harvesting the 10th month crop to get the crops in before the heavy rains. This caused difficulties regarding payments to those auxiliary workers and brigades working for other cooperatives.[32] The productivity of the urban auxiliaries may have been below that of the peasants, and this would have caused resentment if equal workpoints were awarded. The mobilisation of urban workers constituted a direct means of food procurement. This was also the objective behind calls for factories and offices to grow their own food.[33]

The overriding concern for the State remained the relationship between the cooperatives and the other sectors of the economy, in particular the tendency of cooperatives to pursue their own members' interests.

> There still exists a tendency to let rural areas pursue their easy-going and free life

> and to neglect requirements of socialist industrialisation and the lasting interests of the entire people.[34]

Accordingly,

> beginning with the present winter-spring agricultural campaign, the task of mobilising the idle capital of peasants and developing production must be closely related to the task of developing industrial and agricultural production and strengthening all relations between the State and peasants and between the national and collective economy.[35]

The State wanted greater control over agriculture. It believed this could be achieved by amalgamating cooperatives to form hamlet-size ones and educating cadres, specifically with regard to the agriculture-industry relationship, within which

> agriculture must fulfill the principal task entrusted to it by supplying sufficient food to cities, industrial zones and other production bases, and providing a sufficient quantity of raw materials for industrial production.[36]

This relationship incorporated the problems of resource mobilisation, peasant consumption, price and procurement policy, and the planned development of industry.

> Importance must continually be attached to the task of building rational economic relations between cities and rural area. It is necessary to stress continually the fact that objective circumstances require that an exchange of heavy industry products for agricultural products must be organized, not only during the transition phase to socialism but also during the period of completion of socialist building. This is an economic characteristic of the socialist regime. At the same time, it must be clearly understood and continuously remembered that, under the socialist regime, the exchange of goods between industry and agriculture can-

not develop in a spontaneous and disorderly manner as in the past. Continuous emphasis must be placed on the need to take the peasants' consumption of agricultural products into consideration, to organise gradually and plan this consumption, to watch continuously the situation of the purchase of agricultural products, and to carry out boldly and resolutely the purchase policy and task.[37]

Table 5.2: Agricultural Production and Procurement, 1961-65

	1961	1962	1963	1964	1965
Paddy area (m.ha.)	2.39	2.41	2.36	2.43	2.40
Paddy production					
Total (m.tons)	4.39	4.39	4.11	4.43	4.55
5th month (m.tons)	1.71	1.62	1.49	1.77	1.82
10th month (m.tons)	2.56	2.64	2.50	2.53	2.62
Total staple production (m.tons - paddy equiv.)	5.20	5.17	5.01	5.52	5.56
Paddy yield (quintals per ha.)	18.4	18.2	17.4	18.2	19.0
Paddy per capital (kg)	281	256	232	243	240
Staples per capita (kg)	314	301	283	302	294
Paddy procurement (000 tons)	804	890	695	930	1058
Paddy procurement as % of paddy production	18.3	20.3	16.9	21.0	23.3

Sources: Adapted from data in Hanoi (1977c), pp. 117-20 and Vo Nhan Tri (1967), pp. 337 and 451.

Shortcomings over 'ideology' thus consisted of weaknesses over the enforcement of VWP and State policy within which procurement was the priority.

As Table 5.2 shows, food availability was higher in 1961 than in any other year of the Five-Year Plan period.

Some Policy Responses

A report on the situation in one province in early 1962[38] described the major problems of cooperatives. It referred to the only nominal existence of cooperatives. Members devoted more time to working for themselves than for the brigade, with some even engaging in trade rather than collective labour.

It was also noted that,

> few progressive elements who had achieved many successes in production were admitted to the leading organs of basic organisations.[39]

This might be interpreted in two ways: first, few peasants with the ability and belief to develop collective production had attained managerial posts and hence, that these posts had been taken by peasants with ambivalent views towards cooperatives; or alternatively, that successful peasants, in productive terms, had not been admitted to these posts due to their class backgrounds. This could have resulted from rigid adherence to the principle of giving only one-third of management positions to 'middle-middle' and 'upper-middle' peasants. It is quite possible that both situations occurred at different levels. The larger cooperative was more likely to have some higher level VWP involvement in the choice of leadership, whereas the small cooperative or brigade was more subject to tradition.

The 5th Plenum had stated that class background should not be used as a discriminatory criterion for VWP members. It also referred to the emergence of 'a new class--that of cooperative farmers.'[40] In his National Assembly speech in May 1959, Truong Chinh had stressed that classifications fixed during Land Reform had limited usefulness and that the one-third/two-thirds division could be relaxed in certain cases. His references to 'new and old' 'poor' peasants and criticisms of 'class narrowness'[41] indicated the degree of flexibility permissable.

The desirable combination was a leadership (at brigade and cooperative level) consisting of successful producers who believed in the advantages of cooperatives and assiduously fulfilled their duties to the State. According to the report cited above, cooperative management was more frequently characterised by bureaucratisation, conservatism, inefficiency, and corruption.[42] These had adverse consequences for peasant motivation, as well as affecting production directly through, for example, failure to maintain irrigation systems.

In response to the limited organisational abilities at both levels of the cooperative, some sections of the peasantry were reasserting individual family production. This might also have resulted from the peasants ensuring the installation of managements with limited ability, who would not impinge on or limit their autonomy. From the State's perspective, both tendencies could only be overcome by increased control and the installation of competent, committed cadres. In line with this there was an increasing tendency to publicise detailed cultivation and harvesting instructions. This and other points can be illustrated by considering the 'autumn' crop.

Traditionally, the 'autumn' crop had served primarily to prevent a food shortage between the two major harvests. From 1961 it was to be developed on a larger scale and include vegetables, secondary, and industrial crops. In 1961, autumn paddy was cultivated on some 98,000 hectares (almost double that of 1960) and yielded a harvest of 128,000 tons compared with a little under 100,000 tons in 1960. The yield was some 25% lower.[43]

However, the development of a third major crop suffered from a number of difficulties, the most important of which concerned labour organisation. In a relatively short period the peasants

> must harvest the 5th month crop, gather in secondary crops, plough and harrow rice fields for the autumn crop cultivation, and

sow for the 10th month crop.[44]

If manpower were not organised adequately, delays could carry over and affect the 10th month crop, as allegedly occurred in 1961. Other 'shortcomings' of the 1961 crop included inadequate fertilisation and failure to control insects and water. Over 200,000 hectares of the 10th month crop were also affected by water shortages.[45]

Towards the end of May 1962 the Premier's office issued detailed instructions for the autumn crop.[46] They emphasised the need to use all fields not used for the 5th or 10th month crops, as well as fields on which the 5th month crop had been harvested early; to plan and control fertiliser application, weeding, and control of pests;

> to concentrate manpower on the most important agricultural tasks in order to ensure the success of the autumn crop without impairing the 5th and 10th month crops;

and for adequate ideological preparation,

> (cadres) must give cooperatives and their members adequate explanations concerning all tasks--minor and major--take care to avoid compulsion and commandism, and at the same time curb fears of difficulties and negative and conservative concepts.

If ideological exhortations alone proved inadequate, then,

> to stimulate peasants to carry out the autumn agricultural production campaign, the government does not declare it obligatory for paddy obtained from this year's autumn crop to be sold to the State.[47]

Although the 5th month area increased in 1962 to some 930,000 hectares, reported output fell by 100,000 tons to just under 1.62 million tons. The area planted to autumn rice fell marginally below that of 1961, while its output remained constant. Emphasis on the autumn crop diminished in future years as it became clear that it interfered with the two main harvests and had substantially lower yields. The expansion of the area under this crop to more marginal lands reduced the average yield of the crop. The stress on it in 1962 was perhaps a reflection of the optimistic, yet unrealistic expectations of the leadership with respect to all aspects of agricultural production.

The relationship of the autumn crop to the main harvests was not confined to supply-side factors. Peasant labour allocation and hence output (i.e. the actual point of production on the possibility function) was dependent on the marginal return in terms of real consumption to any extra effort expended, assuming basic subsistence had been satisfied. The marginal return depended on the effective terms of trade (incorporating the actual existence of demand for agricultural produce and supply of industrial commodities as well as exchange prices) and the marginal rate of taxation.

If these factors led to a low marginal return, particularly likely in surplus areas away from cities, then there would be little incentive to increase production of exchange crops. The autumn crop would then be a substitute for the main crops or would not be produced. Since its yield was lower, the former would be the likely response. The waiving of quota sales to the State was an attempt to raise the marginal return.

The diversification of production, the raising of the cropping index, and the development of animal husbandry were intended to utilise surplus resources without affecting existing paddy production. It was believed that output from the main harvests could be increased only marginally without major technical developments or increases in the cultivated area, and that unemployed labour would yield greater returns if devoted to new crops. This argument would not necessarily apply to family-based production. The respective production possibility functions can be set out as follows, although the point of production depended on incremental returns as discussed above.

If the marginal product of labour time declined with increases in the labour input, then the cooperative would be unlikely to use labour beyond the point at which the marginal product began to fall (as the value of the workpoint would be reduced). By applying labour to the point where the marginal product was lower, the family could achieve output increases without significant technical progress.

The per hectare 10th month yields achieved in 1959, when agriculture was dominated by family production, were never exceeded--even the 1974 yield was some 200kg per hectare lower. The 5th month yield was exceeded only after the widespread introduction of spring rice strains in the early 1970s. In 1964 the 5th month crop's productivity in a model cooperative was only 5kg per work day--lower than during the colonial period[48] and the 7.5kg per work day recorded in 1959.[49] It apparently remained at approximately 5kg per work day throughout the 1960s.[50] This suggests that even the average product was not higher for collective production.

Fundamental difficulties in the supply of basic foodstuffs, other agricultural products, and consumer goods continued

> due partly to the fact that demand has increased greatly while production remains unable to satisfy it, and partly due to shortcomings in collection, purchase and distribution (which) created a tense situation between supply and demand.[51]

One response to this was food production by factories, army units and offices, and calls for reduction in 'wasteful consumption.' In August 1962 government instructions were issued on 'auxiliary food production.'[52]

> Each organ, school, construction site, factory and army or people's armed police unit must devote any spare time and spare plot to production, in order to ensure that each cadre (worker and soldier) each year produces for his own consumption an average of 10kg of vegetables and more meat (than he is now producing) to improve his diet....It is necessary to make exhaustive use of all available land -

> in gardens, in courtyards, around schools, construction sites, factories and barracks - including land not yet used on construction sites. <u>It is necessary to change the concept about planting gardens in order to eliminate all unnecessary things and replace them by things which are pleasant and useful.</u> (emphasis added)

The instructions called for the use of land unused by cooperatives, the opening up of new lands, the borrowing of means of production and seeds from cooperatives, and finance from trade union funds.

> In animal husbandry, we must make exhaustive use of food refuse and actively produce more food for cattle. It is forbidden to feed cattle with rice, paddy, or fragments of rice grains.

The resulting output would raise individual food availability and consumption and would reduce the quantity which the State supplied.

> Any excess of food and foodstuffs will be bought by the State at encouraging prices. There is no agricultural tax payable on fields cultivated by organs, construction sites, factories, schools, army, and people's armed police units; but where a tax has been paid for the fields concerned in the past, it must continue to be paid by the organs, construction sites...(etc.)...at the same rate.

An emulation movement was launched to stimulate production and convince those involved that the objective was an improvement in their <u>own</u> living standard. Leadership of the movement rested with the Premier's office and extended down to the individual 'production committee.' Labour organisation was to permit adequate time for <u>all</u> tasks, including 'family affairs.'

> They must engage in production in off-duty hours, not in office hours. In cases of emergency - e.g. the need to complete

cultivation on time or to fight against drought or flooding - they may engage in production during office hours, provided they are authorised to do so by their unit head: in such a case, they must work for their units on other days in order to safeguard administrative discipline.

This movement was a response to various problems. First, it illustrated the need to take all possible measures to raise production. It also showed a growing awareness of the difficulty of mobilising agricultural output. Production within the urban sector reduced the need for intersectoral transfers. In addition, the instructions on all aspects of cultivation and production showed greater direct involvement in food production by the central leadership. However, instructions could neither ensure control nor overcome the fundamental constraints. Finally, it illustrated the need for ideological education and intense propaganda campaigns.

> Clear realisation and satisfactory implementation of this policy will not only have an economic effect, but also a direct bearing on the improvement of concepts and views about labour and production, which will result in stimulating everyone to enthusiastic participation in production.[53]

The impact of these measures on production was probably limited and may have adversely affected the normal activities of those involved. The effect on motivation and consciousness is difficult to assess. Weather conditions continued to cause problems in 1962, the technical capabiity of brigades remained seriously limited, and problems of management and organisation were far from solved. Partly in response to difficulties caused by typhoon damage, the following call was made in late September:

> We should strengthen our control over the use of manpower and try to prevent cooperative members from paying attention to their own families' affairs while neglecting the work of the cooperatives.[54]

The Ministry of Agriculture issued instructions on manpower mobilisation at the beginning of the 10th month harvest.

> Special attention must be paid to the very important problem of satisfactory labour management and labour distribution and the drawing up of harvest schedules. Local authorities must see to it that the entire labour force of cooperatives is used rationally.... Leadership over manpower organizations must be closely associated with ideological work throughout the harvest period.[55]

Labour organisation was central to the functioning of the brigade. It involved changing the allocation of labour time between the brigade and family production. It also encompassed the issues of allocation of tasks and distribution of income. Only effective and efficient mobilisation of each brigade's labour force could even attempt to produce the alleged benefits of the larger unit. Inefficiencies and mismanagment would reinforce peasant preference for individual production and make it more difficult to raise collective production. The failure to achieve these increases and the clear managerial difficulties of cooperatives may have led to high-level doubts about the viability of collectivisation given the technical state of agriculture.[56] Le Duan made the following comment in 1962:

> The other day I went to a hamlet where the per capita landholding is only 180 square metres. It then occurred to my mind that <u>given our present poor technical capacities it would be better to let the peasants there till their land individually - cooperativisation would create too many difficulties</u>.[57]

However, the cooperative was viewed as crucial to procurement, which was regarded as essential for rapid industrial development and future 'all-round economic development.' With agricultural production stagnating and per capita paddy production falling, (see Table 5.2) procurement would inevitably become more difficult unless food aid were to rise substantially.

A new purchasing policy, introduced in mid-1962, made more explicit the relationship between the supply of agricultural produce and consumer goods.[58] The 'material reward policy' aspired to encourage cooperatives and individual peasants to raise production, reduce consumption, and sell more produce to the State. The State share of the paddy harvest had risen in 1961 to 18.3%, but in absolute terms had not kept up with the population increase. State purchases of subsidiary crops fell and purchases of meat were over 40% lower, partly because of the falling numbers of livestock (Table 5.1). Meat procurement remained low in 1962 and 1963, and the annual meat ration fell from 6kg in May 1961 to 2.6kg in May 1963. Meat consumption had been over 7kg per capita in 1959.[59] Procurement of food sufficient to satisfy State requirements was a continuing struggle. The proportion obtained through above-quota sales was only 3% of the harvest,[60] while 8% of the harvest was obtained through taxation.[61]

The limited sales of agricultural commodities reflected the limited agricultural surplus and the effective terms of trade. Consumer and intermediate producer goods were in limited availability, relatively high-priced, and often of poor quality.[62] Many consumer items were made by the peasants themselves or by rural handicraft workers.

The policy introduced in 1962 involved increases in official purchase prices for a number of industrial and subsidiary crops: foodstuffs by 3.5%, industrial crops by 5.5%, livestock products by 12.8%, and certain products such as beans and peanuts by 20% to 30%.[63]

> As material rewards, the State may sell, apart from the fixed amounts, an additional quota of production means and consumer goods to units or individuals who have fulfilled, or overfulfilled, the plan for selling paddy to the State.[64]

This aimed at increasing the State's share of production and reducing the scope of the private market. By providing direct incentives of consumer goods, it would limit the peasantry's tendency to consume, or to not actually produce food surpluses.

The provision of these incentives for cooperatives and individual producers was a recognition of the importance of the family sector, which produced a high proportion of subsidiary crops, vegetables, and meat. Moreover, in 1962, 14% of families and 31% of the land were outside the cooperatives.[65] If the State could not induce these families to join and hence gain a greater degree of control of their production, then it would provide incentives to induce them to market a higher proportion of their produce. In 1961, produce grown outside the cooperatives accounted for 65% of the total marketed.[66]

This policy was emphasised during the 1962 10th month harvest, as was auxiliary food production in army units, factories, and schools. An instruction from the Premier's office made clear the three sources by which the State obtained agricultural produce. It also illustrated the desire to limit the price paid for above-quota purchases, the need to limit peasant consumption, and the underlying need to transfer resources to the urban industrial sector.

> It is necessary to endeavour to fulfill the task of collecting taxes and debts and of purchasing food. Concerning the purchase of food over and above norms, the greatest importance must be attached to purchasing an additional quantity of paddy over and above norms and at official prices in areas where this appears to be possible. The aim of this is to increase the amount of food in the hands of the State and to help to solve food difficulties facing the State owing to the need to improve the people's livelihood and requirements of socialist industrialisation. Especially in areas where crops are abundant, peasants must be urged to be economical in the consumption of food in order to be able to sell an additional quantity of paddy.[67]

The objective was to obtain the maximum amount of produce at the lowest cost wherever possible. Only in situations where control was limited or peasants were insufficiently aware of their 'duties to the State' were material incentives to be used.

The instructions quoted above stressed that distribution among cooperators should occur <u>after</u> fulfillment of all duties to the State. Distribution should neither pursue extreme egalitarianism nor be based solely on work contributions.[68] It called for increased control over production, procurement, and distribution; it repeated restrictions on the use of rice for alcohol and animal feeding and it criticised the 'wasteful consumption of food at feasts.'[69]

During the 10th month harvest, calls were repeated for above-quota sales to the State, especially in those areas where the harvest was relatively good. In those areas affected by the bad weather conditions, sales to the State were to be maintained if possible.

> It is necessary to avoid the tendency to magnify losses due to storms and floods and thus reduce the norms for food sales to the State.

In all areas,

> local authorities must pay attention to educating the people to be thrifty with food consumption, give close leadership in this and resolutely combat waste of food.[70]

As Table 5.2 shows, the 1962 10th month harvest exceeded that of 1961 and compensated for the lower 5th month harvest which had been affected by droughts.[71] As the population continued to increase, paddy and staple food production fell. State procurement of paddy rose while that of secondary crops and meat remained constant at a little under 30,000 tons and 60,000 tons respectively.[72]

The extent and spread of cooperatives changed very little during this period. Some 30,000 cooperatives contained 86% of peasant families. Over 10,000 of these were hamlet-size and a few hundred, village-size. The average membership in high-level cooperatives was eighty families. There were also 10,000 high-level cooperatives containing 37% of peasant families, averaging 109 families. Delta cooperatives had, on average, more families than those in the midlands.[73] Production brigades comprised fifteen to twenty families.[74]

Some 225,000 hectares of the 1963 5th month crop were affected by a severe drought. This triggered more criticism of cooperative management and its response to climatic problems. There had already been complaints about badly organised irrigation and flood control work.[75] Peasants displayed a reluctance to partake in this work which was crucial to the 'superiority' of the cooperative. In response, the leadership launched both a major agricultural management and a VWP 'improvement' campaign, believing that these would resolve agriculture's problems.[76]

> And, if this superiority is not already evident to all, the fault lies principaly in the bad management of a certain number of cooperatives.[77]

But the 'certain number' was an understatement. A study of eighteen provinces in late 1964 assessed that in late 1963, only 25 to 30 percent of cooperatives were 'good,' while 52% were merely 'satisfactory,' and 18% 'bad.' By late 1964--some eighteen months after the launching the 'management improvement campaign'--these proportions had changed little to 36%, 48% and 16% respectively.[78]

More significantly, peasant differentiation--a key element in the rationale for collectivisation--was apparently unchanged in 1963. Whereas in 1959, 22% of families had a food surplus, 51% just enough, and 23% were underfed, in 1963 the proportions were 25%, 56% and 19% respectively.[79] Although the figures are probably not directly comparable, they give a rough indication of the persistence of intra-peasantry differentials.

This was attributable to the 15% of the peasantry still officially outside of cooperatives;[80] to the fact that only 39% of peasant income was from the cooperatives;[81] to the dividend on land in low-level cooperatives; and to the differential labour payments in all cooperatives.

The management campaign continued for three years in an attempt to raise production substantially--the targets for the Five-Year Plan period were seven million tons of rice and 9.5 million tons of all staples. If this had been achieved, it would have been possible to raise both peasant consumption and the amount available for exports, industry, and urban consumption. If the marginal propensity to consume staples fell as income rose, then the absolute and relative State share could have risen, given certain conditions.

If production did not increase, then cooperatives under 'improved' management would ensure increased VWP regulation of agricultural output and distribution, and an increased extraction of resources. The stress on procurement was, however, likely to be counter-productive given the backward state of agriculture and consumption levels. A macro level policy based on raising agricultural production by providing appropriate inputs had been rejected in favour of a policy based on rapid industrialisation. The success of this policy was constrained by agriculture and the availability of foreign aid.

A resolution of the Council of Ministers' meeting, which followed the 8th Plenum of the VWP CC, reviewing progress during the previous five years,[82] noted that:

> the majority of our people have not had sufficient rice to eat or clothes to wear.... While the needs of the people and the State are thus increasing daily, agricultural production is every year threatened by natural calamities. Each poor harvest brings with it new difficulties.[83]

There were two causes for this. First,

> we did not pay sufficient attention to sup-

plying means of production, consumer goods, and so forth, to the peasants, with a view to encouraging them to produce and sell more produce to the State.[84]

And second, there was a failure to 'understand fully the needs of the State' and a failure 'to grasp fully the connection between immediate and future interests.' The inabiity of the peasantry to understand that the building of socialism required short-term sacrifices had caused 'excess' rural food consumption, private trading, and high prices. The response was a modified mixture of policies.

In addition to the attempts to increase direct State control and the 'education' and 'economy' campaigns, incentive policies were also readjusted in 1963. Sales quotas for cooperatives were fixed over a longer period in the hope that they would guarantee State supplies and stimulate increased production.

> The quota of food to be sold to the State will be fixed in terms of absolute weight, in accordance with the actual production situation, the consumption requirements of the cooperatives and the peasants, and the average quantity of food sold to the State in the past three years (1960, 1961 and 1962). The quota of food sold to the State will be readjusted in favour of agricultural cooperatives and individual peasants in the event of their production having undergone a drastic change or of their harvest being very poor.[85]

Above-quota production could be sold to the State at prices up to 50% higher than those paid for quota sales. The differential between quota sales and above-quota sales had risen to 50% from 10%-20% in 1960, and 30% in 1962.[86] In late 1963, the prices were 0.40 _dong_ per kilo for above-quota sales--equal to the retail price of processed rice, and hence representing a loss for the trading organ--and 0.27 _dong_ per kilo for quota sales.[87] The agricultural tax--paid in kind and averaging 7.2% of the harvest--was also fixed in 1963 on the same basis for the remainder of the

Five-Year Plan period.[88] The policy was changed by 1965 when quota sales again varied in accordance with the crop.[89]

Means of production, building materials, and consumer goods were set aside for peasants who sold produce at these incentive prices.[90] Successful implementation of the policy depended crucially on the ability both to provide these goods and to maintain incentive prices. Other crop prices were also raised in 1963 by up to 50%. Drought had affected certain industrial crops in particular, severely reducing peasant cash income and demand for industrial products in places.[91]

Adverse weather conditions in 1963 caused a fall in output and low per-hectare yield of 5th month paddy and affected the 10th month crop. Some areas suffered almost continual drought followed by water-logging and sudden floods.[92] Total paddy production fell by 280,000 tons in 1963 with per capita production falling below 240kg for the first time since 1954. The State share fell by 195,000 tons--70% of the fall in total production--indicating the extremely limited food surplus and the proportionately greater consequences for procurement of harvest fluctuations. An effect of the fall in output was a rise in the free-market rice price in 1963; this would have diverted some production from official channels.

A change in the structure of trade in 1962-63 may also have contributed to State purchasing problems. In 1962 supply and marketing cooperatives were mostly merged with the enterprises controlled by the Ministry of Internal Commerce,[93] and their share of retail trade fell from 17% in 1962 to 6% in 1963. The reversal of this change in subsequent years--by 1966 the supply and marketing cooperatives' share was over 18%[94]--may reflect peasant opposition and reduced trading.

The Technical Development of Agriculture

The various campaigns continued in 1964. The agricultural situation improved somewhat but the commencement of US bombing raids caused increasing concern. Greater attention was focussed on developing the extremely backward technical base of agriculture. The key non-labour inputs to agriculture were water, fertiliser and seeds.

There were two reasons to focus on irrigation and water control. First, the potential for mass labour mobilisation was arguably greater with co-operatives. Second, given the backward state of agriculture, returns to this would be proportionately greatest.[95] It would permit more intensive use of existing land and an extension of the cultivated area. There were, however, limits to increasing the cropping index due to soil exhaustion.[96] In the 1950s, land reclamation had concentrated on rehabilitating the 140,000 hectares destroyed during the war with France--a task which was completed by 1958.[97]

Table 5.3: Investment in Agriculture and Irrigation, 1955-71

Year	Agriculture Investment as % of Total Investment	Irrigation Investment as % of Agriculture Investment
1955	16.2	91
1956	13.9	81
1957	11.0	79
1958	15.2	73
1959	11.5	44
1960	10.7	41
1961	16.2	45
1962	21.3	48
1963	23.4	48
1964	23.0	64
1965	....	66
1965-68	14.0	62
1969-71	21.0	64
1972	....	62

Sources: Official sources cited in Vo Nhan Tri (1967), pp. 532-42; Le Chau (1966), p. 202; Nguyen Tien Hung (1977), p. 123; and Nguyen Xuan Lai & Vu Quoc Tuan (1976), p. 161.

The bulk of investment in agriculture (Table 5.3) was devoted to irrigation during this period.

It produced an increase in the irrigation network and an increase in the cultivated area. The Five-Year Plan target was to reclaim 550,000 hectares, although the actual achievement was somewhat less than this. The establishment of new cooperatives in the midland and highland areas involved relocating peasants and introducing new crops and cultivation techniques. They thus took time to become established and produce a marketable surplus.[98] However, given the extent of the irrigation network developed with traditional techniques used under the pre-colonial and colonial regimes, there were limits to what could be achieved without substantial new inputs. There were also problems in motivating the peasants to participate, particularly in maintenance. The specialist irrigation teams developed from 1963 onward were an attempt to ensure adequate and frequent maintenance.

Fertiliser utilisation reflected its availability, which was mainly from human and animal manure. Chemical fertiliser was in extremely limited supply throughout the period.[99] A frequently quoted target was for the application of ten tons of fertiliser per hectare.[100] Achieved levels reportedly rose from 2.7 tons in 1955 to 3.6 tons in 1957, 5.8 tons in 1961, and 5.7 tons in 1963.[101]

The failure to achieve the target reflected both its ambitious level and also livestock problems. Peasants were exhorted to use other sources including 'azolla,' cultivated as 'green manure,'[102] pond mud, and alluvial silt. Campaigns to increase human manure collection were also introduced.

The data on yields (Table 5.2) suggest limited seed development despite reports of new strains being developed, based mainly on imported Chinese varieties.[103] Specialist livestock breeding was also based on imported strains.[104]

The development of agricultural implements focussed on improving existing ones rather than on mechanisation. Improved tools would complement an increased labour input and increase output without major capital inputs. Mechanisation was confined to the 2% of land cultivated by the State farms of which there were fifty-eight in 1963. These farms suffered problems of organisation and motivation in a more attenuated form than cooperatives and failed to fulfill their intended role as the most progressive part of the agrarian sector.[105] Ambitious plans for their expansion were shelved in the mid-1960s, and emphasis was focussed on cooperatives.[106]

An early, over-ambitious plan was to introduce mechanisation on 15% of the cultivated area by 1965, in addition to the State farms.[107] By 1964 it was accepted that 'for a long time to come our agriculture will essentially depend upon draught animal power, handicraft labour, and improved and semi-mechanised tools.'[108] However, the implements produced by the industrial sector were frequently criticised for being of poor quality, inappropriate, and in limited supply,[109] while the rural handicraft sector had been run down prematurely.[110] Maintaining collectively owned implements and organising their 'efficient' utilisation also caused problems. Specialist teams within cooperatives were later organised for the purpose of repairing and 'improving' tools.[111]

Agricultural 'research' during this period appears to have been limited to issuing detailed crop instructions, guidelines, and timetables. However, the widespread dissemination of existing knowledge and 'best-practice' techniques did achieve a certain amount of success.[112] Limited resources were available for agriculture because of the industrialisation priority. Therefore, attention focussed on measures which increased output at low resource cost. This involved mobilising and motivating labour within the framework of collective agriculture.

Shifts in Focus, 1964-65

Average paddy production during the years 1960-64 was below that of the previous four years, although subsidiary crop production rose. The claimed increases in the number of work days[113] thus confirmed the declining marginal productivity of labour time in the absence of increased technical and capital inputs to production, while the problem of collective labour organisation may have induced a desire to shift to greater capital intensity. The belief that collectivisation could _substitute_ for new techniques by way of a more sophisticated division of labour and larger scale mobilisation, at best, remained unproven.

Le Duan had asserted in 1962 that 'if six and a half million peasants work not only 130 days but 200 in a year, and besides, if the value of a working days rises from one _dong_ to two _dongs_ for example, then it is clear that the social labour productivity will increase sharply without any technique.'[114]

The relative failure of collectivisation to raise production or labour productivity had made the maintenance of the family sector an economic necessity. However, it was possible to argue that the maintenance and tolerance of this sector had _inhibited_ development and that 'fully socialist' cooperatives would demonstrate the superiority of the collective form.[115] This was a frequently used argument from late 1963 onward.[116] An alternative, equally mechanistic interpretation of the dominant ideology could lead to the opposite conclusion. Le Duan referred to cadres, who, after reading theories in course textbooks which said that industrialisation must precede collectivisation, had doubted the logic of VWP policy.[117]

This latter view does not appear to have been expressed at the highest level of the VWP. Here there dominated a belief in both industrialisation and its requirements from agriculture, and the 'superiority' of cooperatives. Leadership differences were concerned with how to establish this 'superiority' and 'tax' agriculture rather than with its underlying logic. They differed also over whether family production complemented or conflicted with collective production, and over the economic and political consequences of its removal. The policy outcome reflected a leadership compromise as well as one between the peasantry--within which there were differences--and the VWP.

The management campaign continued through 1964 and was cited as contributing to the improved 5th month harvest.[118] However, the favourable weather conditions and 1963 procurement policy changes were probably more important contributory factors. Exhortations after this harvest not to indulge in 'excess consumption' and not to neglect other crop preparations illustrate the State preoccupation with procurement. There was concern that peasants would substitute rice for subsidiary crops in consumption and thus reduce production of the latter. The off-take of all crops might then not increase. Production of subsidiary crops had doubled between 1960 and 1964. Their cultivation and consumption were repeatedly emphasised because they reduced risk. They also employed under-utilised resources and could help relieve the food shortage. However, the population preferred to maintain rice consumption if possible. It was also claimed at this time that livestock production had been neglected.[119] Cooperatives were told to

> carry out distribution satisfactorily within cooperatives, save funds, meet food obligations, achieve economy, and resolutely fight the tendency to indulge in continual festivities on the pretext of celebrating successful harvests.[120]

These tendencies reflected forced substitution or forced saving. Peasants raised their own consumption because of exchange prices, limited supply, and poor quality of industrial goods.[121] The same factors provided little incentive for accumulation. The level of cooperative accumulation was below the five to ten percent envisaged in the Statutes,[122] while the potential for accumulation within the family sector was limited. The VWP saw these as further signs of the peasantry remaining 'under the manifold influences of the ideas and habits generated by the small peasant economy.'[123]

Nevertheless, State off-take of paddy rose in 1964 by a factor of 0.7 times the change in output. Subsidiary crop and meat procurement also rose considerably over the low levels of 1962. The traded share of agricultural production, in value terms, rose to 18.3%, the highest level achieved during the 1960s.[124] The stabilisation of tax and purchasing policies had undoubtedly improved the supply of produce to the State. First, the compulsory quota no longer fluctuated with the harvest and this resulted in higher above-quota sales in 1964,[125] and hence a rise in peasant income. Second, intermediate and consumption goods were reserved 'for use in a sales-reward system for units and individuals who have distinguished themselves in selling goods to the State.'[126] The rise in the share of consumer goods in total imports from 10% in 1962[127] to 22.2% in 1964[128] may have contributed to this policy, as would the increased per capita production (Fig. 6). Increased availability of consumer goods could lead to a decrease in hoarding by peasants with surplus food.

The need to link agriculture more firmly to the urban industrial sector by way of the mutual exchange of goods was considered of paramount importance. Successful development of the 'sales-reward system' during a period of relative abundance could stimulate output and trade. A substantial <u>potential</u> demand already existed which would become <u>effective</u> if peasants produced more. However, at existing output levels, per capita productivity and income were declining and the marginal propensity to consume food (from necessity) was extremely high. Satisfaction of peasant demand could create a desire to maintain consumption of these goods, even in more constrained periods, thereby raising the marketed <u>share</u>. In addition, peasants would increase <u>output</u> in order to raise purchases. One constraint on agriculture was the supply of consumer goods.

In addition to the management campaign, there were emulation movements (notably one based around the Dai Phuong cooperative in Quang Binh province) and a movement to develop agricultural implements which focussed on the establishment of specialist cooperative irrigation teams.[129] These teams consisted of twenty members specialising in irrigation tasks for two to three months each year.[130] Most of their work was remedial or small-scale. Major irrigation projects required greater resources than the cooperatives had at their disposal.

Estimates of the extent of the irrigation network vary considerably. The basic irrigation and water-control network was constructed on the foundation of the one already existing during the 1950s which provided service to 60-70% of the cultivated area. Irrigation investment then fell as a proportion of all agricultural investment (Table 5.3). Further major developments may have faced a technical and resource constraint. Moreover, beyond this point the marginal return to further investment expenditure might fall or be higher in other activities. Further small-scale advances were made and irrigation was extended to subsidiary and industrial crop areas in the 1960s. However, the major network required considerable maintenance.

Criticisms of the irrigation campaign focussed on its leadership for failing to coordinate activities and stimulate enthusiastic peasant participation.[131] The peasants were not paid for this work and they lacked the direct incentive of having their <u>own</u> land involved, as had been the case in the late 1950s. However, the rise in the cropping index from 1.62 in 1963 to 1.67 in 1964 was attributed to irrigation developments. This index had stood at 1.58 in 1960 with a target of 2.0 by 1965.[132] An additional irrigation campaign was launched in 1966-67 with the same objectives--80% of the rice area, 50% of the subsidiary crop area, and 60% of the industrial crop area.[133] This reflects the limited success of the previous campaign and the impact of US bombing of the network.

In later years, particularly during the severe 1971 floods, there were criticisms of the inadequacy and under-utilisation of the alleged capacity of the irrigation network. In addition, in 1964, as the southern conflict escalated and placed increasing demands on the resources of the DRV, a 'one works as two' campaign was launched.[134] These campaigns shared the objective of maximising resource use by raising the 'ideological level' of the population and its commitment to socialism as conceived by the VWP leadership. They recognised both the economy's limited technical capacity and that the collective form had not yet inculcated a 'socialist consciousness' upon the peasantry. They also showed the influence of developments in China, notably, the 'Socialist Education Movement.'

> Cooperatives do not become progressive merely because they enjoy especially favourable conditions: they become progressive because, in these cooperatives, the thoughts of the masses have undergone an advantageous change and the cadres and masses have the determination to overcome difficulties in order to advance firmly and to produce more material wealth and, thus, to improve their living conditions and, at the same time, satisfactorily fulfill their duty towards the State. All this contributes to achieving socialist industrialisation in the North and concretely supporting the southern compatriot's struggle for self-liberation.[135]

Truong Chinh referred to the conservative outlook of cadres as a main reason for the slow growth of agriculture. He reactivated the ideological debate, referring to the 'struggle between the two paths' and the 'struggle against rightist thoughts.' He called on VWP cadres to organise, lead, and set examples for the rest of the peasantry. His comments reflect the belief that <u>many</u> of the problems of the agricultural sector resulted from inadequate leadership and the failure of cadres to act as examplars.

This 'failure' of cadres has a number of explanations. It could have been the result of inadequate technical education, or it could have been due to conflicts between the cooperative level cadres, brigade leadership, and families. It might reflect 'small-scale social corruption'; the struggle to improve family conditions could have lead to cadres making use of their position. Finally, it might indicate a lack of conviction of the advantages of cooperatives, which could reflect a more general discontent. In terms of successful policy implementation, this 'failure' was particularly serious since VWP membership in most rural areas was extremely low. It also indicates the limited success of attempts beginning in 1958-59 to educate cadres ideologically.[136]

In 1963 VWP membership in many provinces constituted a mere 1-2% of the population.[137] Cadres recruited during the resistance still constituted 70% of the membership, while those recruited during collectivisation were predominantly young people. The strength of the middle peasants who had joined during the resistance and exerted their influence in 1956, thus remained strong. Moreover, VWP cadres had not taken a major role in cooperative management, even in those cases where a VWP cell existed. A 1964 report showing that 50% of VWP members remained 'unconvinced' by the collectivisation arguments, indicates the problem for the VWP leadership. It meant that there were, at most, a handful of committed VWP members in each cooperative and even less, if any, in each brigade! Even if one assumes that VWP cadres were representative of the rural population, at least half of the peasantry misunderstood 'the importance of building and developing the base of a collective economy.'[138] Presumably they favoured a family-based economy.

The 10th Plenum of the 3rd VWP CC in December 1964 repeated the orthodox view that intersectoral relationships and State control of them needed developing; that agriculture had to provide food, raw materials, and export products (the performance of the latter had been partially disappointing); and that the three revolutions (in technique, production relations, and ideology) were all important in fulfilling these objectives.[139] The policy response on which it focussed was the development of a 'rational' and comprehensive price system, 'serving and stimulating the vigorous and balanced development of industrial and agricultural production.'[140] This would ensure that peasants and cooperatives sold the maximum amount of produce to the State, constrained their own consumption, and received intermediate goods essential to the expansion of production.[141]

However, in terms of <u>concrete</u> policies, the resolution contained nothing new and did not even explicitly recognise the fundamental constraints on achieving the above objectives, desirable as they might be. It represented a further attempt to exhort cadres to fulfill their duties, improve organisation, and in turn, exhort the peasants.

In 1965 the campaign to 'improve' cooperative management was developed and expanded. In accordance with the focus on control of agriculture, it emphasised the raising of low-level cooperatives to high-level ones and hence, the abrogation of rent payments; the grouping of cooperatives to form larger ones; and the incorporation of families and land still outside the cooperatives.[142] The southern conflict may have added momentum to these developments, although the more direct impact, in terms of bomb destruction, labour mobilisation, and industrial decentralisation came later. These emphases repeat the arguments given earlier for collectivisation that larger high-level cooperatives could reduce inter-peasant differentials, benefit from increasing returns to scale, and aid procurement. The rise in output and marketings in 1964 may have permitted a renewed emphasis on 'consolidation' of cooperatives by relieving pressure somewhat.

Opinion within the VWP leadership may also have shifted for a number of reasons. First was the belief in the necessity of further developing the cooperative form (it was 'socialist'). Second, there was evidence that cooperatives had only nominal existence in many areas; and third, there was concern over dependence on material incentives to individual producers. Also adding to the changed opinion was the clear indication that the potential for technical development of agriculture was extremely limited without a decisive shift away from the heavy industrialisation strategy or substantial aid.

Thus it was argued, as it had been in 1958-59, that cooperative production could substitute for these developments by more efficient use of land and labour. The effect of the policy shift on the family sector was clear:

> Efforts must also be made to settle the problem of management of the cooperatives land and ricefields satisfactorily, and to put an end to the use of the cooperative ricefields for personal purposes, to provide guidance for the effective use of

> the 5% of ricefields and land reserved for cooperators' families.[143]

Further, it was necessary to re-emphasise the relationship between the two sectors:

> The secondary family economy is primarily designed to make good and rational use of the spare time of cooperators.[144]

The official argument was that limitation of the family sector could allow more labour to be devoted to brigade production which, better organised, would yield production and productivity increases. It would also limit differentiation and concentrate output in the hands of fewer, larger, cooperatives under greater VWP control.

> Efforts must be exerted to urge cooperators to devote about 250 work days to cooperative production and to increase the number of workdays per hectare of paddy from 300 to 400 on each crop.[145]

At this time peasants devoted, on average, 150 to 160 workdays per year to the brigade.[146] and 200 in the 'best socialist cooperatives.'[147] The value of a workday remained below one <u>dong</u>, indicating no increase in average labour productivity since the early 1960s, assuming constant relative prices and State exactions. As suggested earlier, the marginal physical productivity of labour time may have been declining. The number of work days per hectare had been 333 in 1959.[148] Increases in agricultural output since then resulted from an increased cultivated area rather than from higher yields per hectare.[149]

The output of subsidiary crops approximately doubled between 1960 and 1965. However, maize yields remained constant, potato yields rose by 15%, and manioc yields declined by 20%.[150] A potential for higher yields existed but could be achieved only with increased inputs and incentives. The increase in diversification into industrial and subsidiary crops may have contributed to the stagnation of paddy production by its demand on resources of labour, water, fertiliser, and collective management. Relative incentives also affected the crop mix.

Average data for workdays conceal differences related to the differentiation argument. Peasants with more resources in the form of capital, livestock, and implements devoted more labour time to their family plots.[151] Depending on labour resources, they would also have accepted increased privatisation of usufruct. Given that on average over 60% of income derived from non-cooperative production,[152] it is not surprising that intra-peasant differentials persisted. The membership of the different strata continued to change over time because of intra-peasant socio-economic mobility.

By 1963, 30% of 'poor' peasant families had raised their income to that of 'upper-middle' peasants, while there remained a 30% difference in the average income of these two groups.[153] The 2nd Five-Year Plan objective, emphasised at the 5th Plenum of the 2nd VWP CC in July 1961, had been to raise the living standard of all peasants to that of 'upper-middle' peasants.[154] Differential family earnings from brigade work were due partly to differences in family size--a study in 1964 found that 'poor' peasant families had 1.58 adult workers per family and 'middle' peasant families 2.13.[155] Differences in family size were related to past living standards and demographic factors. In 1965 all cooperative families averaged 1.87 adult workers. Over the whole period between 1961 and 1965, average peasant family income rose by 13.4%, while the gross value of agricultural output at constant prices rose by 17.5%.[156]

Thus, in 1965 differentiation comparable to that of 1958-59 still existed, procurement remained problematic, and cooperatives had failed as yet to demonstrate any material or technical superiority. Agricultural output had risen more slowly than population, exacerbating the procurement problem and reducing per capita food production. Food imports were very low in these years and hence, per capita production equalled per capita availability in any given year, assuming no stocks.

In the years following Land Reform (1956 to 1959) staple food production averaged 335kg per year and some food was exported. During the Five-Year Plan period from 1961 to 1965 it averaged 299kg.[157] Vegetable and bean production had almost doubled but remained low, while meat availability had fallen because of a fall in livestock numbers. Moreover, the escalating military conflict increased the need for control over food resources and for consumption economy. A Nhan Dan editorial in November 1965 called for a reduction in the daily rice ration (currently around 13kg a month for an adult) and a reduction in the above-quota rice price.[158] This illustrated both the war pressures and the continuing vacillation over the use of material incentives. It is not clear that the price was reduced, but paddy production did fall in 1966.

Development of high-level 'socialist' cooperatives could be rationalised in 1965 by using the same arguments as in 1959, although with no more, and possibly less, supporting evidence. The view that the case was unproven because of a failure to 'consolidate' cooperatives ignores the reasons why this had not occurred.

By the end of 1965 the percentage of peasant families in all cooperatives had risen to 90%.[159] The proportion in high-level cooperatives rose from 45.5% to 65% in 1964. More significantly in terms of control of resources, the amount of cultivated land outside cooperatives fell from over 30% in 1963 to below 20% in 1965. In 1964, 7.5% of the cultivated area was brought in, while the proportion of families remained constant. This indicates that during this period, and in 1964 in particular, the rich peasants joined cooperatives. It also indicates the extension of collectivisation to the high-land areas where average landholdings were higher than in the delta areas. However, even in 1965, the 10% of families outside coopertives held 20% of the land while high-level cooperatives held proportionately less land than families. The average size of cooperatives remained largely unchanged during the years 1961-1965, although it rose considerably during subsequent years.

There are a number of explanations for rich peasants joining cooperatives. There was increasing pressure on them and on the family economy, although discriminatory pricing and input availability had existed for some time.[160] After the good 1964 harvests and the changes in State purchasing policies, the free-market rice price had fallen to the level of the official price, making marketing independence less attractive. Furthermore, they realised that their interests would be better served by being inside, rather than outside, cooperatives. The potential for cooperatives to protect peasant interests vis-a-vis the State has already been emphasised, and will be discussed further in the context of increased village and rural autonomy during the war with the United States.

Summary

This chapter has summarised the development of agriculture during the period from collectivisation in 1959-60 to the beginning of US bombing in 1964-65. It illustrated the contradictions of DRV agricultural policy and the continued search for a combination of policies to resolve these contradictions.

The chapter began with an outline of the nature of agricultural organisation, stressing the potential conflicts of interest between the State, cooperatives, and peasants, and also within these groups. Examples of these conflicts were presented. They tend to counter the 'official' view of the complementary nature of collective and household production.

The livestock problems of 1961 showed the peasantry's response to the attempt to collectivise livestock. The also revealed the failure of the VWP to understand the peasantry and predict its response, as well as the close relationship of livestock to grain production in the DRV. Falling grain output resulted in less food being available for livestock and also in a short-term substitution of meat for grain. Thus, livestock numbers fell, reducing fertiliser availability (there was no chemical fertiliser available) with consequent effects on grain output.

The various policy responses considered in this chapter illustrate the VWP awareness of the contradiction between collective form and reality. They also show the carrot-and-stick approach to the peasantry, and the intractable nature of the dilemma. Direct procurement and urban auxiliary production were attempts to overcome supply problems--the former affecting work incentives, the latter, urban production. Higher-level and larger cooperatives and emphasis on 'improved management' were attempts to gain greater control over production and output and to overcome peasant ambivalence directly. Ideological campaigns attempted to produce the same effect by changing peasant 'consciousness' and inducing them to accept their duty to the State and consume less. Incentive policies appealed directly to material interests.

Peasant decision-making was considered in the context of these policies by focussing on labour allocation. It was stressed that a considerable number of factors entered this equation, but that the higher productivity of household production as against collective production, suggested by both empirical and non-empirical evidence, was a major constraint on the successful operation of cooperatives. It also reduced any incentive for non-members to join, while making the State dependent on sales of private producers and hence on provision of certain material incentives.

It was also stressed that different peasant households had different attitudes towards labour allocation. This depended on the same range of factors which influenced their overall attitude towards collectives as discussed in Chapter IV. The ambivalent attitude of cadres was also emphasised, as were the major problems this posed for the central leadership, since rural cadres were relatively limited in number.

The extremely limited technical development of agriculture was considered briefly to illustrate a number of points. The limited assistance provided by industry and the focus on 'traditional' inputs--irrigation and organic fertiliser--derived in part from a belief that organisation (institutional change) could raise agricultural production with existing technology. This had, of course, been one aspect of the argument for collectives. It also reflected the industrial bias of DRV policy--agriculture would have to wait for assistance--and a certain ignorance regarding agriculture. Other evidence in addition to that from the DRV suggests that the return to labour-intensive irrigation works may have been limited, at least after the point already attained in the DRV.[161] The attempt to develop autumn paddy must be regarded as a failure since it had low yields and competed with existing crops. Some positive results may have been achieved, however, through the widespread dissemination of existing 'best-practice' techniques.

There is almost no evidence that, amid this shuffling of policies, consideration was given to clear alternatives such as increased industrial and State assistance to private agriculture. Private production was tolerated as a short-term necessary evil as it has been in the USSR and China.[162]

The remainder of the chapter discussed the year 1964--a good year in many respects. It was argued that the incentive policies may have been a major contributing factor to the improvement. Throughout this work, emphasis has been placed on the constraint posed on agriculture by the limited supply of consumer goods. It was argued that the improved situation permitted the leadership to launch another drive forward in collectivisation. In doing so, it employed the same three arguments used in 1959, although there was no new evidence to validate them. The vacillation over the use of incentives was again illustrated. The changed circumstances were also given as a reason for a number of land-rich peasants entering the collectives.

CHAPTER VI

Agriculture in a War Economy

Managing a War Economy

From 1965 to 1975 the DRV existed under a war economy. The decade subdivides into two periods of major US bombing--1965 to 1968 and 1971 to 1973, and two periods without--1968 to 1970 and after the Paris agreements of 1973. The impact on production during these periods is shown in Figures 1 to 8.

For industry, the war involved a shift to war-related production, although substantial quantities of armaments were imported, mainly from the USSR. Aid from China consisted largely of food and consumer goods. Domestic production of consumer goods would be expected to decline in relative importance. The war also required moving industrial production, wherever possible, to locations where it was less exposed. The greater likelihood for destruction of heavy industry acted as a counter for the tendency of relative output shares to shift in its favour. Relocation of industrial production eased problems caused by damage to transport and communication systems as well as the prior demands placed on them by military-related activities.

The effect of war on the labour force was to divert labour to the military and defence tasks, although much of the latter was on a part-time basis. Given the labour surplus in agriculture and rural areas in general, mobilisation was unlikely to affect production directly, although the problem of supplying those who had been mobilised still remained. This increased the procurement pressure on agriculture. The war may have diverted skilled technical labour and cadres to war-related tasks which might have affected collective production.

This shift of industrial production to local areas would stimulate local demand for labour and for agricultural production. If accompanied by increasing consumer-good production it might stimulate agriculture and the marketed surplus. However, the shift seems to have involved a relocation of heavy industry rather than of consumer goods, with any change in shares being a result of proportionately greater heavy industry destruction. Handicraft production continued to stagnate (see Figures 3 to 6). The ability to provide production inputs to agriculture was increasingly constrained.

Despite increased foreign assistance, the rise in government expenditure had budgetary consequences given the supply constraint facing the economy. Open inflation appeared in 1965 and 1966. Since agriculture, by way of trade, provided the major source of government revenue, the pressure on the terms of trade it faced increased. That the economy actually kept going throughout the war is partly a result of the dominance of agriculture and labour-intensive production which is much less susceptible to direct damage. It is also attributable to the people and the leadership.

The increased emphasis on decentralisation and local self-sufficiency, as well as the damage to transport and communications, and the cadre shortage, all contributed to increased village and cooperative autonomy. This combined with an expanding private market--particularly in the early 1970s--may have stimulated agriculture in certain cases, either on the basis of family production or in efficiently organised cooperatives with a favourable resource base.

Decentralisation Under Siege Economy Conditions

The escalation of the southern conflict and the US bombing which had begun in 1964 exacerbated the difficulties of the DRV agrarian sector. Hoang Anh attributed the weaknesses of agriculture as follows:

> Partly because the material and technical bases of our agriculture are still weak, and partly because of difficulties caused by the enemy's war of destruction. <u>But the main reason is unsatisfactory leadership</u>.[1]

He stressed the need for 'collective leadership' and criticised the 'old production habits of a backward agriculture.' Despite these and other injunctions, many cooperatives continued to have a nominal existence and the family economy remained significant.

State procurement of agricultural production remained problematic. The war increased the need to mobilise output for the urban labour force, for defence workers, for industrial raw materials, and for export.[2] The supply of industrial goods in exchange was affected by direct damage and the shift to wartime production. However, the decentralisation of industry could increase local availability of consumer goods.

Price and procurement policy received major attention, as did calls for economy in consumption. One measure adopted was the prohibition of 'illegal' distillation of alcohol.

> The illicit distilling of alcohol wastes a considerable quantity of foodstuffs and causes difficulties for the labour management of cooperatives. Moreover, it may lead to other social vices such as gambling, black-marketeering, disunity and so forth.[3]

Exhortations against the use of rice to make alcohol had been issued on previous occasions. The use of law suggests that 'education' and persuasion had had limited effect. Given the longstanding opposition to colonial edicts on alcohol distillation and use, circumvention of the exhortations and law are not surprising. State trading branch tasks for 1966 included the following:

> to accelerate the collection and purchase of goods; to keep a firm hold on the source of goods; to guide the people in saving commodities; to improve distribution; to strengthen the management of markets; and to strive to stabilise prices and currency.[4]

Although the State sector officially accounted

for 90% of retail trade at this time,[5] not all inflationary pressures were suppressed. The official cost-of-living index for classified workers, which had been relatively stable since 1961, rose by 7.5% in 1966 and 6.4% in 1967.[6] Real income of classified workers declined by 26% between 1960 and 1971.[7] The impact on the peasantry depended on free market price changes, while their purchase of consumer goods were constrained more by supply than price. However, the shortages of both agricultural and industrial commodities, along with the increased budgetary expenditures, clearly contributed to the open inflation of these years. The State apparently lowered 'the prices of production materials for agriculture and the prices of a number of consumer goods'[8]--again with budgetary consequences.

As Table 5.2 shows, paddy procurement reached record levels in 1965. Subsequent poor harvests coupled with a continuing population increase put great pressure on consumption levels and made procurement inevitably more difficult. The traded share of the value of agricultural production fell from 19.9% in 1965 to 13.8% in 1970, while retail sales of staples were maintained, presumably because of increased imports.[9]

The relatively good 1964 and 1965 harvests may have resulted in a build-up of reserves which were run down in later years.[10] The claim that compulsory deliveries averaged 20% of production during the war years[11] seems implausible without either increased sales to the peasantry or extreme pressure on rural living standards. The State had the ability to pursue neither of these policies. If the 20% refers to total procurement then it may reflect either under-reporting of output which would tend to rise as per-capita consumption fell, or a greater willingness to deliver since peasant family members were employed in the army or on war work. Food imports, amounting to between 500,000 and one million tons per year during the second half of the 1960s, contributed to price stabilisation and mitigated any further pressure on peasant consumption. US estimates of imports suggest they averaged 15% of food eaten and 11% of rice during the war years.[12]

The underlying problem remained the inability of paddy production to keep pace with the population increase. From 1964 onwards, it was below the level required to provide all the population with the consumption standard of 240kg per year. Total staples production was below this level in 1968, 1969, 1971, and 1973 and only barely above it in 1970 and 1972. Table 6.1 summarises production data over the period 1966-75.

Larger, high-level cooperatives, which were intended to alleviate the problems of production, procurement, and rural differentiation, continued to be developed. But problems of management and motivation remained. Official data on cooperatives did not necessarily accord with the situation in reality.[13] The average data also conceal differences between provinces, villages, and cooperatives.[14] Differences in performance also tended to increase over time[15] and were given added impetus by the decentralisation and local-level autonomy during the war years.

The most significant shift in policy during the early war years was the localisation of industrial production. This had two aspects: the decentralisation of key industry and the encouragement of local industry in the form of small-scale production, both modern and artisan. It was partly justified on the grounds that it 'was appropriate to the stage of development of the economy' and 'would aid the transformation of the economy and the transition to socialism.'

In 1965 local industries' share of the total value of industrial and handicraft production in the DRV had fallen to 46%.[16] Investment in local industry was 3.5 times higher in the years 1965-67 than it had been in 1962-64.[17] Its share rose to 55% in 1967, partly as a result of the increased investment, the relocation of certain industries, and the destruction of much centralised heavy industry.[18] Its share of all industrial and handicraft production had fallen from 43% in 1960 to 27% in 1965.[19]

The official rationalisation of local industry development was that it would fully utilise local resources (labour and capital) and hence raise

Table 6.1: Agricultural Production, 1966-75

	1966	1967	1968	1969	1970	1971	1972	1973	1974	1975
Paddy area (m. ha.)	2.39	2.19	2.08	2.15	2.21	2.07	2.20	2.09	2.27	2.25
Paddy production										
Total (m. tons)	4.13	4.29	3.71	3.91	4.46	4.12	4.92	4.47	5.49	4.78
5th month (m. tons)	1.46	1.67	1.62	1.59	1.78	2.19	2.25	2.15	2.58	2.11
10th month (m. tons)	2.53	2.56	2.04	2.27	2.62	1.91	2.64	2.29	2.87	2.63
Total staple production (m. tons paddy equivalent)	5.10	5.40	4.63	4.71	5.28	4.92	5.74	5.19	6.28	5.50
Paddy yield (quintals per ha.)	17.3	19.6	17.8	18.2	20.2	19.9	22.4	21.4	24.2	21.2
Paddy per capita (kg)	212	214	179	184	204	184	215	190	228	194
Staples per capita (kg)	262	269	224	222	242	220	250	221	261	224

Source: Adapted from data in Hanoi (1977c).

accumulation. It would stimulate heavy industry to serve the needs of local industry, itself more orientated to agricultural needs, and it would create 'local equilibria between industry and agriculture, production and distribution, and circulation and consumption.' It was presented not as a temporary measure but one consistent with the overall objective of socialist transition.[20]

The argument was thus based on the need for increased self-sufficiency and the need to move key industries to make them less susceptible to destructive attacks. It would also alleviate the increasing difficulties with transport and communications. Further, local industry could aid the accumulation process by providing consumer goods and local demand for agricultural output. Less desirable from the State's viewpoint, self-reliance increased the ability of local levels to circumvent policy guidelines. The basic strategy remained 'to use heavy industry as the basis: to give priority to rational development of heavy industry.'[21] The rise in the share of local industry in the total resulted from the relocation of heavy industry rather than from accelerated development of consumer goods or handicrafts production whose shares and growth rates were little changed.[22]

Agricultural policy continued to emphasise the maximisation of output from existing resources. This was one reason from completing high-level collectivisation. It required increasing the labour time devoted to the brigade, the workdays per hectare, and fertiliser collection and use. It also involved reparceling of fields, crop specialisation, and labour-intensive irrigation work. As before, these policies encountered problems stemming from family allocation of labour time and the possibility that increased allocation of labour to brigade production reduced the value of the workday and workpoint. Peasant opposition to partaking of collective accumulation, organisational difficulties, and the lack of incentive provided by real exchange possibilities were additional problems.

Rationalisation of field layout, intended to assist intensive cultivation as well as irrigation and drainage, had apparently extended to 38% of the cultivated area by 1967.[23] Continued emphasis[24] is perhaps indicative of the peasantry's attachment to the post-Land Reform holdings, collectivised or not. Fertiliser availability was constrained by the slow growth of animal husbandry,[25] still largely a family occupation.

It was claimed in early 1966 that 83% of the rice area, 32% of subsidiary crops, and 34% of industrial crops were irrigated, and that cooperatives had established 15,000 water conservancy teams.[26] However, only one-third of the irrigated area was watered 'according to scientific methods.'[27] Its quality and that of the teams were questioned during the late 1966 drought.

> According to the evaluation of the water conservancy branch, 80% of the total cultivated area can be protected from drought by irrigation projects. But what is the reason for the shortage of water on 400,000 hectares of ricefields of the 10th month crop (i.e. one-third of the total area) during the time the drought was at its peak? It is true that we have built many anti-drought projects, but we must manage and exploit them properly and, in particular, must secure a perfect network of small irrigation projects so that we can develop completely the efficiency of all projects and the rate of efficiency of these projects will not be really too inferior to the planned capacity.[28]

Despite a further two-year irrigation campaign launched in 1966, similar comments were made in future years. The 'form' of extensive irrigation systems and teams did not accord with their functioning in reality.

Similar problems emerged with other aspects of production. Success in increasing the cultivated area in places resulted in a stagnation or fall of productivity per hectare. This was partly because of the extension of cultivation to more marginal lands, but also because a given amount of brigade labour was spread over a greater area. With labour utilisation, increases in the labour input did not always produce commensurate increases in food output because of the low marginal product of labour time on existing land with existing inputs. The problem also emerged in reverse. Campaigns to produce five tons per cultivated hectare achieved favourable results in places, but at the expense of reductions in the cultivated area.

> Various localities have attached much importance to intensive cultivation aimed at achieving a paddy yield of 5 tons per hectare while not devoting adequate attention to the complete cultivation of the planned area. As a result they have achieved the desired yield but have not increased their total output or have permitted it to decline. Some areas, in their scramble to win the 'Five-Ton' title, have sought to abandon ricefields with unstable output, thus showing complete disregard of total output.[29]

This illustrates a divergence between brigade and 'social' evaluation. The State wished to maximise total output by making maximum use of the labour force, although it also wanted to see average labour productivity increase. Average labour productivity rose during the war because the agricultural labour force fell by one million between 1965 and 1968.[30] The brigade, by concentrating effort on the best land, endeavoured to maximise labour productivity and the value of the work-point. If the marginal productivity of labour time was falling as labour time increased, the output resulting from the brigade decision would be lower than that desired by the State.

The alleged complementarity of the family sector would not affect this divergence if it devoted surplus labour time to the family plot rather than to the brigade. Output maximisation could, in theory, result from either the use of <u>all</u> labour time by the brigade and the overcoming of any leisure/work preference of all production. Peasant family decision-making, however, would be heavily influenced by the evaluation of work and leisure, itself dependent upon the commodities available for exchange and their prices.

Output maximisation remained the fundamental State concern. It needed food to feed and increase the urban workforce, as well as other agricultural resources in the form of raw materials and exports. The 'priority to heavy industry' strategy required resources for accumulation from agriculture in the absence of sufficient aid. A means of ensuring this creation and transfer of resources was thus a priority for the State. Even if State industry appeared to be self-financing through its own profits, accumulation still had to be realised by means of a price system for industrial products manufactured by State enterprises and sold to peasants. The strategy was based on two tenets: that the industrial sector could transform resources more efficiently than agriculture (i.e. the potential for productivity increases was greater)[31], and that the advantages of squeezing agriculture outweighed the disadvantages. The limited surplus available for accumulation aggravated the problems of implementing this strategy.

A speech by Deputy Premier Pham Hung in early 1967 shows appreciation for this dilemma.

> The needs in various fields are very great and have rapidly increased, but our capability to meet those needs is limited.[32]

The problems of reconciling demands for consumption and accumulation (one result of which was inflation) were caused by

> the weak material and technical bases of socialism; the very low output of social labour--especially the labour output in agriculture; and, the small increase in national income and socialist surplus production, partially damaged by the war.[33]

He argued that regardless of long-term solutions, in the short-term the conflict could be mitigated by improving distribution systems, reducing consumption needs to a minimum, maximising labour utilisation, and encouraging a degree of self-sufficiency on the part of schools, factories, and offices. This had been emphasised earlier (see

Chapter V) and was aimed at cultivating unused land, using labour which could not be transferred elsewhere but was still under-utilised and, most important, at bypassing trade. His argument was that the potential did exist for solving the problem of competing demands, at least partially. However, its realisation was inhibited by managerial, organisational, and distributional flaws, as well as by a failure to inculcate 'socialist consciousness' into workers and peasants.

Emphasis on the latter was a substitute for the provision of material incentives and aimed at restricting consumption.

> Our people and cadres have a very high revolutionary spirit. Our people are not afraid to live in poverty and to suffer a shortage of food and clothing. What they ask for is a fair and rational distribution method within the country's capacity and conditions, and consistent with the distribution policy of the Party and the State....Along with improving the distribution methods, we must urge the people to practice thrift in consumption and to buy only necessary goods, in order to avoid an artificial shortage.[34]

This article focussed on control over the <u>sources</u> of supply of goods which was facilitated by collectivisation. Its objectives included the encouragement of self-sufficiency, consumption economy, and a system of distribution designed to penalise those who did not work to their best potential. This latter point is linked to the 'drive' to raise the utilisation of labour by encouraging longer hours and more workdays to the extent that during the 1967 10th month harvest

> the most important thing is to encourage cooperative members to prove their devotion to duty by working both day and night.[35]

These analyses correctly pinpointed the problem, but believed its solution was greater control over labour, production, and consumption in agriculture. Better organised and consolidated cooperatives should maximise output and procurement. The drive to establish larger high-level cooperatives thus continued until 1968. It appears to have involved the incorporation of the larger peasant landowners, particularly in 1966.

Agricultural output statistics suggest that the belief in cooperatives was misplaced. Production of basic foodstuffs was at best stagnant and falling in per capita terms. Vegetable and bean production, although rising, could make only minimal impact on overall consumption levels. With regard to animal husbandry:

> The quantity and quality of the livestock have remained poor and the number of cattle has continually decreased in the past few years. Therefore, the ever growing need for draught animals and farmyard manure to meet the requirements of intensive cultivation and crop increases: the need for foodstuffs, such as meat, eggs and milk, and the need for animal husbandry products in industry and export have not yet been satisfied.[36]

Considerable emphasis was placed on developing animal husbandry and pigs in particular. Specialist breeding was carried out using imported stock, an additional 5% of cooperative land was set aside for feedstuff cultivation, extension services were provided, and price incentives introduced. However, problems continued with pig breeding; veterinary care was inadequate and, more important was the shortage of feed. Pig-raising has remained largely a family occupation and the State has had difficulty persuading peasants to sell pigs rather than eat them. This reflects both the real exchange problem and the desirability of meat consumption.[37]

The Old Approach Reinstated, 1968-70

The 1968 bombing halt again changed conditions for the DRV. It also allowed consideration of the consequences of the war, the effects of policy shifts, and the prospects for the future. The increased focus on local development was confirmed, while the need to maintain a balance between excessive centralism and excessive regionalism was noted.[38] The degree of autonomy which had developed at various levels had resulted in an increased exercise of power by bureaucratic or corrupt officials. Attention was called to the need to control and restrict corruption, embezzlement, and 'tendencies to commandism.'[39] Petty corruption became more likely as the economic situation deteriorated. This was particularly true at the lower levels where the managerial and technical ability was lower, the pressure greater, and rewards minimal.[40]

Concern with productivity, consumption economy, control of prices and markets, and the need to improve the management and organisation of agricultural production of course continued. The falls in agricultural and industrial output, especially pronounced in per capita terms (see Figures 1 to 7), inevitably made price and procurement problems more difficult. The divergence between form and reality was noted with respect to agricultural implements.

> While the requirements for agricultural implements have not yet been fully met, millions of implements of various types have accumulated, undistributed, in storage. The organisational task, from production to circulation, distribution and use of implements, has not yet been satisfactorily fulfilled.[41]

This was partly an inevitable consequence of local production, self-sufficiency, use of available materials, the relatively low technical level of production, and the demands placed on skilled cadres by the war. However, similar criticisms were made earlier and repeated in later years. By 1968, high-level cooperatives officially included 88% of peasant families; only 5% of families and 8% of the land remained completely outside of cooperatives. However, these percentages do not reflect the extent of collective production of even basic foodstuffs, nor the degree to which cooperatives were able to organise, plan, and integrate all agrarian activities. Family production, constrained in principle, had in practice been accorded a greater degree of tolerance. The

State and VWP had to avoid alienating the peasantry including cadres who zealously guarded their right to family production. Moreover, the absence of significant output increases, and hence secure procurement from the cooperatives, made the State dependent on produce from the family sector.

The increased extent of nominal high-level collectivisation meant that cooperatives represented almost all the peasants in any particular area and acted as an intermediary between the peasants and the State. Local-level autonomy had also risen during the war years because of decentralisation, self-sufficiency, and the disruptions to transport and communications. The concentration of both political and economic power in the hands of village cooperative and brigade cadres, increasingly independent of the State, provided opportunities for petty corruption and the pursuit of particular interests. The alleged practice of cooperatives (and presumably brigades) keeping three sets of account books--the cadres' version, one for members, and one for officials from outside[42]--illustrates the potential degree of autonomy and cadre acquisitiveness.[43] The fall in food production per capita during the war years may have been exaggerated by an increased tendency to under-report as conditions worsened, autonomy increased, and cooperatives included all the peasants in a locality. These same factors may have led to a reinstatement of the historical character of the village-commune as a semi-autarkic entity.[44]

In 1968, as a response to these tendencies, a further debate developed on the need to make collective production a reality.[45] As before, the argument was that cooperatives (under greater central control),[46] could raise output via returns to scale, could limit differentiation both within and between cooperatives, and could increase State control over distribution. The disastrous 1968 harvests, which probably reduced staple food availability below the 240kg level, certainly stimulated the third argument. It was also plausible to argue that the material superiority of cooperatives had not been established precisely because they had not been effectively 'consolidated.'

Criticisms were thus levelled at the limited utilisation of labour by the brigade. This was central to raising brigade production and deriving the scale benefits. The peasantry allocated their labour time between their own activities (which included non-agricultural production) and the brigade, presumably based on a rational conception of the best way of satisfying their material needs given exchange possibilities. The consequence was that insufficient labour was devoted to the brigade to maximise brigade output within existing technical constraints.

As long as the marginal product of labour time on brigade land is positive, then the allocation of more labour raises output. However, the marginal return to the peasant may be greater by allocating that labour to the family plot or even to leisure, if the income is insufficient reward or there is nothing on which to spend the money. In addition, labour was not deployed on other developments. On average, around 150 workdays consisting of four to five hours, were devoted to collective production. Within these averages there were wide discrepancies and workdays were not always devoted to the brigade at the most appropriate time. Work on the family plot tended to receive priority.

> Recently, although the crop was ripe, a number of cooperative members continued to busy themselves with their trade or with planting vegetables on their own plots of land. Cadres were unable to mobilise many people to go harvesting. This situation is mainly due to the fact that concrete measures have not been taken to settle correctly the relationship between the cooperative and family economy.[47]

The thrust of the argument against the family economy was that it constrained collective cultivation and concentrated output in the family where its mobilisation was less direct than in the cooperative. The criticism was not merely that disproportionate amounts of labour were applied to the 5% plot, but that this minimum land allocation was widely violated in practice.

> A number of localities are erroneously applying the three-condition contract system to families--especially families engaging in cultivating secondary crops, in transplanting or in animal husbandry. In other localities cooperatives distribute their land to members' families and allot one or several production tasks to them for a season or a longer period. Or, again, cooperative members' families are allocated a certain quantity of land and, at harvest time, they must hand over to the cooperative a certain quantity of products. In other cases cooperative members' families are allotted some land which they cultivate, enjoying as much as they produce.[48]

One or more of these practices apparently applied to rice cultivation in at least 30% of cooperatives, while in some provinces deviations were much more widespread.[49] Cultivation of subsidiary and industrial crops as well as vegetables and livestock husbandry were predominantly family occupations, either on the 5% plot or subcontracted from the cooperative or brigade. The 'three-contracts with families,' although involving the transfer of temporary usufruct and not ownership, were severely condemned by Truong Chinh.

> The three contracts with families are in reality a return to the individualistic way of working. They destroy the meaning of the agricultural cooperative movement and make cooperatives a mere form.[50]

Truong Chinh's analysis was remarkably similar to that which he had presented some ten years previously.[51] He argued, at some length as usual, that cooperatives facilitated both the development of new techniques and the better utilisation of existing ones. Irrigation, fertiliser use, and the division of labour could benefit from the returns to scale inherent in the larger organisations. The introduction of modern equipment required the peasants to be assisted by trained personnel which was easier within cooperatives. Again, he alleged that family production 'creates an economy of small commodity production: this economy every day and every hour gives rise to capitalism.'[52]

This latter argument related not merely to the 'three-contracts with families' practice, but to the 5% plot itself. Truong Chinh thus explicitly recognised the contradictory aspects of the family-economy/collective-economy relationaship in the existing conditions, rather than the complementary aspects stressed in other presentations.[53]

This re-examination of agrarian developments and repetition of the collectivisation arguments culminated in the Statute on High-Level Cooperatives, promulgated in April and published in August 1969.[59] This Statute and other articles of the period emphasised management and organisation as the major deficiencies. They were the features which could validate the argument[55] and increase State control. They were also cited as the main reasons why success had not already been achieved. Failure did not lead to a reconsideration of the whole strategy of collective agriculture, but rather to a renewed emphasis on bigger, better-organised cooperatives.

In some sense the leadership was reasserting an article of faith: that cooperatives were socialist and hence superior to other forms of organisation. This reflected an unjustified confidence in the ability (and honesty) of cadres, as well as a disparaging view of the peasantry. 'It is necessary...to overcome the negative spirit of the old way of working and of the small, individual, dispersed, local, narrow-minded and conservative production system.'[56] The following is a further example of this attitude:

> The technical revolution of 'underdeveloped' agriculture must be adjusted to...the peasants' low cultural level and small capacity to assimilate techniques.[57]

The largely theoretical conception of collective agriculture ignored a number of crucial constraints which included the shortage of able, trained cadres and the opposition of the peasantry to collective production. This opposition was clearly shared by many cadres and was the reason why production had devolved to, or actually remained at, the level of the family. A fundamental constraint, and one apparently least questioned, was the failure of cooperatives to realise the alleged economies of scale. The following comment from a Nhan Dan editorial of the period applies just as plausibly to the central leadership.

> A few branches and establishments formulated plans bureaucratically, separating themselves from reality and from the masses.[58]

The 5% plot, although now actually owned by the cooperative, reamined sacrosant in the Statute. In practice the Statute appears to have produced little change:[59] production continued to take place at the level of the family, and this provided the major portion of peasant income--Table 6.2. The material incentive policies remained in force from economic necessity, and the free market continued to be tolerated.[60] A similar re-examination in 1974 resulted also in a new Cooperatives Statute.[61]

Table 6.2: Peasant Income, 1965-75

		Source of Income[1]		
Year	Index[2]	Cooperative	Family Production	Other
1965	113.4	39.2	51.8	9.0
1971	135.6	35.3	53.2	11.5
1973	147.9	33.9	53.0	13.1
1974	162.4	36.0	52.6	11.4
1975	161.4	34.8	54.3	10.9

Notes: 1. The prices used for valuing the components are not known.
2. 1961=100

Source: Hanoi (1977c) Table 137, p. 180.

One continuing aim of these developments was to raise labour productivity. Manpower was not abundant during the war: there were no unemployed workers, but labour was under-utilised. There was potential for raising the number of workhours per day and workdays per year. This could free labour time for other purposes such as irrigation work or accumulation. One objective was to reduce the number of workers per hectare from three or four, to one, by increasing the number of workhours from five or six, to eight, and the number of workdays per worker to 250 or more per year. This suggests maintaining the number of workhours per hectare, per year, constant at approximately 2,000. However, given existing techniques, increases in land productivity to the target of at least five tons per hectare required disproportionate increases in labour input due to the declining marginal productivity of labour time. This did not apply in all cases to all activities: for example, more careful weeding might not have a declining marginal product over a certain range.

Labour utilisation in agriculture was highly seasonal and the appearance of unemployed labour over the course of a year masked labour shortages at particular times.[62] In 1960, Truong Chinh had referred to a rural labour 'surplus' of 50%.[63] This was claimed to have fallen to 30% in 1962 and to between 30-40% in 1964.[64] A report from 1970 referred to the scope for raising 'the aggregate duration of productive labour by 50%.'[65] At harvest times calls were frequently made to raise both the number of workhours per day and the quality of the labour.

> In the past, during times of pressure many cooperatives could mobilise only 70-75% of their workers, who worked less than 8 hours per day and had a low standard of efficiency. This year cooperatives must mobilise at least 90% of the workers, who must work at least 8 hours per day.[66]

However, at other times of the year there was little work to be done in rice production. Thus it was implausible to ensure that the annual requirement of workhours be provided by one worker.

What was required was the provision of alternative employment possibilities <u>within</u> the rural sector either in current agricultural production (i.e. diversification), capital construction, or handicraft activities. Workers could then spend part of the year engaged in these occupations.

It was estimated at this time that 100kg of paddy required on average 15-16 workdays, and that each peasant cultivated only 0.53 hectares.[67] This average labour productivity of 6.5kg per workday compares with other estimates of 5kg per workday for cooperatives, 7.5kg per workday in 1959,[68] and later estimates of 8kg, 13kg, and 14kg for three 'good' cooperatives.[69] Generalisations and averages mask substantial differences between different cooperatives and brigades. Two types of cooperatives worked quite, or even very well. They were some of a small number of 'model' cooperatives and those cooperatives with well-organised autonomous brigades in surplus areas, particularly close to cities.

The target of five tons and one worker per hectare suggests a target of 20kg per workday if 250 days were worked, and 17kg if 300 days were worked, assuming constant average productivity per workhour. The data above referred to 1.9 workers per hectare. Thus 'surplus' labour consisted of hours and days at particular times of the year for more than one worker. In the absence of developments which could balance the labour demand throughout the year, this surplus was not easily transferable to other sectors. Technical changes within agriculture such as weeding, fertiliser application, or crop diversification could change the seasonal labour pattern by raising labour use.

Its use within the cooperative involved the management and organisation of a whole range of supplementary activities by both the cooperative and brigade. This put a further strain on managerial resources in many cooperatives. The family economy might make greater use of the labour in staple food production by increasing the labour input up to the point at which the marginal physical product was much lower than for cooperatives. It would also receive the incremental return, whereas this might not be distributed by the brigade or cooperative. Further, it was not dependent on the labour of others and was thus more controllable and suffered less from organisational problems. The high proportion of income derived from sources additional to the cooperative (see Table 6.2) reflects these considerations.[70]

Raising labour productivity was also a concern in heavy industry.[71] The DRV's development strategy was based on a belief that industry, particularly heavy industry, was the major source of productivity increases. There is no reason to doubt that the following remained an appropriate description:

> Machines are being utilised to only 50% of their capacity, and a worker works an average of only 21 days a month and 60% of the workday. If the rate of utilisation of machines were raised to 70% or 80% and if, by a strengthening of discipline and the improvement of food, lodging, social services, and working conditions, the monthly average per worker could be raised to 23 or 24 days, and if 80% or 90% of the workday were effectively utilised, productivity would increase substantially.[72]

The proposed solutions were the same as for agriculture--firmer control and better organisation, _and_ material incentives.[73] The former encountered cadre shortages and peasant (or worker) opposition, while the latter encountered the supply scarcities of _both_ agricultural and industrial commodities which severly constrained the ability of either sector to provide incentives to the other.

An important resolution on 'grain' policy was passed in March 1970.[74] Its objectives were to raise total production and increase national and regional level self-sufficiency; to improve the distribution and mobilisation of labour within cooperatives; to encourage consumption economy by measures including control of private trade; and to improve inter-sectoral trade to raise funds for accumulation and the continuing conflict. It recognised that both cooperatives and purchase and distributing policies had, in many places, developed independently of State guidelines. Existing policies were not acting as a satisfactory 'lever' to stimulate the development of agriculture--of food output in particular.

One reason cited for the lack of incentive was the peasants' fear

> that the more they boosted production, the greater would be their contributions to the State.[75]

In practice, as was also noted, output might have risen but was concealed by the peasants, brigades, and cooperatives. Le Thanh Nghi asserted that this fear was 'without reason,' although experience had evidently provided sufficient reason for it. Tax and quota sales had fluctuated between a fixed quantity and a proportion of the harvest. He also referred to other tendencies which illustrated the conflict between the State and peasantry.

> Fraudulent reporting on the area under cultivation, production, output, population, lending ricefields, wasting ricefields, reducing the cultivated acreage of the cooperative, increasing the cultivated acreage of the family economy.[76]

In order to allay peasant concern:

> For 1970, the State has fixed the quota of grain--paddy or maize--to be sold by the cooperatives and peasants and to be credited towards taxes, purchases or collections. Other taxes will be paid in cash. This quota will be in force for five years, from 1970 to the end of 1974. During this time the quota will be changed only when major changes occur in the area planted to paddy and maize by cooperatives.[77]

The quota was calculated on the basis of the most abundant, poorest, and average harvests reaped between 1963 and 1969, with allowances for seriously grain-deficient areas. It could be adjusted in the event of major natural calamities only if no other crops could be cultivated. Exceptions were made for areas specialising in secondary crops and for the highland regions. After fulfillment of the quota, surplus grain could be used for livestock, sold to the State, or sold to those deficient in grain. To ensure a degree of stability and regularity:

> The State will not increase the fixed grain sale quotas of the cooperatives, peasants' associations, and localities that accelerate production, meet their fixed State grain sale norms, and produce large amounts of grain.

Equally,

> Cooperatives, families or localities which produce little must eat little. The State will not sell them food to compensate for this deficiency.[78]

A consequence of the policy would be increased differentials between regions, villages, and cooperatives. The State hoped that encouraging self-sufficiency and fixing quotas for five years would stimulate low output areas to raise production and productivity to achieve basic consumption levels. The high output producers would be able to realise in their present and future consumption levels any output increases, assuming that goods were available to exchange for surplus production. This related to the fifth objective of the new policy concerning inter-sectoral exchanges.

An additional focus of the policy was on income distribution within the cooperative. This was to be similar to income distribution between areas.

> To encourage the cooperative members who have worked well and devoted much of their strength and labour to their cooperatives, and to educate those who are lazy and unconcerned with the building up of cooperatives, it is necessary to achieve distribution of food supplies

> inside the cooperatives according to the principle of distribution in relation to labour: the more one works, the more one is rewarded: the less one works, the less one is rewarded; healthy people who do not work can enjoy nothing.[79]

There were special provisions for soldiers' families, the old and sick, and certain key cadres who had insufficient time to engage in directly productive labour. The latter group could buy grain from other cooperative members, but as the emphasis was on minimising 'unproductive' labour throughout the economy, cadres were expected to engage in productive labour wherever possible. The indication was that such a policy had not been adhered to, and as a result 'the eagerness of workers to produce was not stimulated or developed.'[80] In some areas, distribution had tended to be based on extreme egalitarianism or arbitrariness rather than on strict recording, evaluation, and calculation of workpoints. The guidelines for this were clearly laid down but they involved problems of supervision and work assessment as well as imposing demands on the ability and honesty of cadres. Brigade-level distribution was also criticised.

The importance of 'fair' distributions within the cooperative was recognised, although the following quotation could be adapted to apply to peasant/State distribution.

> Distribution depends on production: but, good or bad, distribution in turn affects the trend of production, the level of fulfillment of the duty towards the State, the capital accumulation of the cooperatives and the cooperative members' livelihood and eagerness to work.[81]

Another objective of the 'food grain policy' was to improve government control over that portion of production not retained by the cooperatives and peasants. Self-sufficiency was intended to achieve consumption economy in certain areas, while the usual general strictures were issued. The main concern was to ensure that surplus producers marketed their excess production via

State or cooperative trading agencies, or directly to those experiencing shortages. They were to be discouraged from selling to private traders, and merchants were prohibited from dealing in grain and related products. Le Thanh Nghi accepted, however, that the free market could not be legislated away.[82]

In practice, private retail trade which had held some 8-10% of the market since 1960, increased its share in 1970 to 18%. The absolute value of State trade fell in that year. The share of private trade continued to rise to 21.5% in 1974.[83] Differential inflation rates may mean that the changes in absolute volume were less dramatic,[84] but they still illustrate the divergence between theory and practice.[85]

The State was consistently unable to obtain significant above-quota sales due to the differential between the State above-quota price and free market prices. From 1965 to 1968 the sale price of rice on the free market was reportedly three times the State price, suggesting that the purchase price difference was also considerable.[86] In earlier years quota purchases accounted for 85% of all State purchases.[87]

The 'grain' policy was also concerned with developing inter-sectoral exchange. It recognised three requirements: to extract the maximum resources from agriculture in food, exports, and industrial crops; to provide sufficient industrial goods to stimulate production and marketings; and to retain sufficient output production and resources within the urban industrial sector for accumulation. However, the resolution itself said little about how these were to be reconciled apart from emphasising that inter-sectoral exchange needed to be under State control.

Le Thanh Nghi emphasised the first requirement--the 'duty of the peasantry'--and inveighed against a pro-peasant position.

> It is necessary to avoid the tendency to unilaterally emphasise the peasants' interests while neglecting the State's needs or to take an interest only in

> purchasing grain which the peasants have a duty to sell while neglecting to purchase it beyond their duty. This tendency does not help the State solve its difficulties concerning the food problem, but results in causing more difficulties on the grain market and further complicating the relations of free trade of grain in the rural areas.[88]

This remark and the continuing calls for consumption economy should be seen in the context of the extreme pressure on food supplies deriving from the series of poor harvests during these years (see Table 6.1).

The policy was aimed at stabilising fixed-quota grain sales in the interests of both State and peasantry, and the distribution of grain between sectors, regions, and individual cooperators. The overall objectives remained raising food production and productivity, and the economy's ability to satisfy the consumption needs and aspirations of workers and peasants in addition to accumulation. In case of conflict, the latter was the priority.[89] The policy was implemented alongside a campaign to develop the cooperative members' 'consciousness of their collective mastership' and one which opposed private production. The latter shared many of the above objectives and was stressed in a VWP CC resolution.[90]

These campaigns were also related to the 'popularisation' and 'implementation' of the new Cooperatives Statute of 1969 which aimed at making collective production a reality. Although problems were partly attributed to leadership failings at all levels, the major criticism was levelled at the peasantry and it echoed the disparaging view mentioned earlier.

> Because our country was previously a colonial and semi-feudal country with a backward economy, we have encountered difficulties in managing State affairs and the Socialist economy in advancing directly towards socialism. Furthermore, the influence of feudalistic ideology,

> the backward customs and habits of the old society and the thoughts and habits of old producers have contributed to a great extent to the above situation.[91]

The policy remained unquestioned; it was the implementers and participants who were at fault! Hence, further cooperative 'improvement' campaigns were necessary.

> Everything must be reviewed again and again in two or three years and checked and complemented every year before it can go along the right track in rural areas.[92]

The period of re-examination of agrarian policy and practice (from 1968 to 1970), thus produced a reaffirmation of principle with emphasis on making cooperatives a reality while curtailing the family economy, private trade, petty corruption, and bureaucratisation. It produced the new Statute on cooperatives and further admonitions to raise the level of socialist consciousness, fulfill duties to the State, and curtail consumption (by humans and animals).[93] In addition, attempts were made to impose stricter central control of agriculture. However, the new Statute had also produced a reaffirmation of practice. The degree of privatisation of production was shown to have been much higher than officially sanctioned and would remain so despite the new Statute.

Private trade had maintained and increased its share despite State opposition in principle. More significantly, the material incentives policy was still considered necessary to obtain sufficient agricultural produce. The State showed a preference for use of the stick to cajole a 'backward, conservative' peasantry, 'irrationally' unconvinced of the benefits of cooperatives. It was, however, necessary to use the carrot of incentive prices and goods to entice peasants to trade produce from both the family and collective economies.

Development in Adversity - 1970-73

Heavy US bombing recommenced in late 1970.

From then until the signing of the Paris agreements in January 1973, the focus of attention returned to the military conflict which now included the whole of Indochina. The policy emphasis remained unchanged: a better organisation of existing resources could satisfy the demands of consumption, accumulation, and war.

> Now it is all the more necessary to strengthen and improve management in order to make the most rational use of the abundant labour force of our people and the material force still largely left unused. This consists in making the most economic and thrifty use of both the means of production and consumer goods whereby constantly to increase social labour productivity and increase social products, increase the national income and the economic potential of North Vietnam, the firm basis for our growing fighting force against US aggression, for national salvation.[94]

With two reasonably good paddy crops, agricultural output in 1970 was better than it had been for some years. Favourable conditions continued during the first half of 1971 before major flooding caused a serious deterioration. This required considerable inter-regional and foreign assistance, the cultivation of alternative crops, as well as State amortisation of some cooperatives' debts.[95]

One technical development that produced positive results was the cultivation of 'spring' rice during the winter-spring cultivation period. This crop is sown in February-March and harvested in June, as opposed to the standard fifth-month crop which is sown in October-November and harvested in May-June. It had been made possible by seed development using imported, high-yielding varieties and local strains.[96] It took advantage of 'dry season' weather conditions and also left a period during the winter, between tenth-month crop harvesting and spring crop sowing, when a third non-rice crop could be cultivated. In 1971, for the first time, the spring rice area amounted to more than half of the total winter-spring area as shown in Table 6.3.

However, the intended benefit from cultivation of crops between the tenth-month and spring harvests was not realised. The area and output of subsidiary crops (maize, manioc, and potatoes) fell from the 1967 peaks of 657,000 hectares and 1 million tons of paddy equivalent, to 470,000 hectares and 722,000 tons in 1973.[97] This fall can be attributed to the reduction in available labour resulting from the war. Other contributing factors included the lack of incentive to cultivate secondary crops; the peasantry's aversion to their cultivation and consumption (good rice harvests often resulted in reductions of subsidiary crop cultivation);[98] and the increased labour input required by the new higher-yielding rice varieties.

One difficulty in raising the cropping index was the need to raise output at minimum extra total cost. If production expenses are high partly because of low labour productivity, then although peasant income rises) it may not appear commensurate to the effort expended, and may result in the peasants carrying out cultivation 'perfunctorily.' The tendency for cooperators to neglect the quality of work in order to complete more and gain more workpoints had been noted earlier.[99] It was necessary to encourage the utilisation of resources so that the incremental return more than justified incremental expenditures. The dilemma was to raise production by intensive resource use while maintaining, and if possible increasing, both land and labour productivity. This reinforced the need for adequate organisation and work incentives. Longer term effects on land productivity were also a concern. 'We must not worsen soil conditions... for the sake of temporary interests in a crop season.'[100] The priority task for the State, however, was to increase total output and hence it placed continued stress on the use of all land, however marginal, and on auxiliary food production.

> Secondary crops must be grown on all suitable categories of land - from collectively owned to family gardens,

Table 6.3: Spring Rice, 1965-75

Year	Area sown with new strains (ha.)	As % of total 5th/Spring area	5th/Spring yield (quintals per ha.)	5th/Spring output (m. tons)
1965	3,700	0.4	18.86	1.821
1966	23,692	2.5	15.22	1.460
1967	25,054	2.8	18.94	1.667
1968	34,500	3.9	18.27	1.617
1969	63,243	7.1	17.82	1.595
1970	103,644	11.5	19.79	1.781
1971	540,000	57.5	23.33	2.193
1972	650,000	68.3	23.66	2.251
1973	....	....	22.33	2.152
1974	....	....	27.33	2.580
1975	800,000 (target)	80.5	21.18	2.106

Sources: Nguyen Xuan Lai and Vu Quoc Tuan (1976), pp. 163-64; Hanoi (1977c) Tables 79-81, pp. 118-20.

> canal banks, roadsides, and any pieces of cultivable land.[101]

As has been suggested repeatedly, the maximisation of output objective was more likely to be fulfilled by the family economy than by the collective economy in the economic and technical conditions of the DRV at the time. However, the shortage of industrial goods constrained both. The quality of consumer goods, agricultural implements, water control systems--the latter particularly after the severe 1971 flooding--and the development of the highland regions all attracted attention during this period. Criticism of the difference between the existence and utilisation of the water control and irrigation network became increasingly severe. In September 1972, Nhan Dan had this to say about a number of major hydraulic systems, the key link in the proposed development of a comprehensive water control network.

> (The) exploitation of these projects is still low in view of their capacity.

> They are unable to meet the requirements for development, and do not justify the cost and labour invested. The reason for this is that these projects are not yet perfect and still present many shortcomings in project planning, construction planning, management and use of water.[102]

The 1971 flood reduced the tenth-month harvest below two million tons. This was partly compensated for in 1972 when, with favourable weather conditions and the expansion of spring rice, paddy output reportedly reached its highest level since 1959 at just under five million tons. Total staples production nearly surpassed the 1959 level at 5.7 million tons. However, population had risen by almost 50% since 1959 and reported staple food production was barely above the 240kg level per capita. Industrial crop production[103] and livestock husbandry also developed favourably in 1972, although meat availability remained below the levels of the late 1950s and early 1960s.[104]

Planning for Peace-Time Agriculture

The signing of the Paris agreements in 1973 again transformed the environment in which the DRV was developing. It allowed a further reconsideration of development trends and progress. This process culminated in a conference held in August 1974 in Thai Binh province which laid down agricultural policy for the foreseeable future. This conference and the accompanying VWP CC directive on agriculture and new Cooperative Statute drew together many of the themes which gave the entire period its element of continuity and have been the basis of much of the preceding analysis.

The peasants' conception of material interests and the failure of the majority of cooperatives and brigades to demonstrate relative material superiority resulted in a 'disproportionate' allocation of labour time to family activities. Brigade work was also frequently of poor quality.[105]

> When our people realise that every work-hour brings them an income and that the better they work the more prosperous they will become, they will enthusiastically

> do better and better work on more and more days. As a result, the income obtained from the collective economy will be increasingly greater than that from the family's side-line economy and this will end the situation in which people do their work negligently.[106]

Failure to realise this vision continued to be attributed to inadequate management both in practice and in existing guidelines.

> Because the rural situation has changed basically, methods of management cannot remain unchanged. It is necessary to ensure that the guidelines for a cooperative comprising 40 or 50 families - in which work is done by primitive means and rice cultivation is the main occupation - are also consistent with the management of a cooperative comprising 300 or 400 families in which there are many branches and occupations and machines are used.[107]

The outcome was the New Management System (NMS) which strengthened and improved guidelines. Emphasis was placed on the district as the appropriate unit of rural organisation, 'ideological education' was continued, and efforts were made to improve the technical base of agriculture. Cooperative superiority was used as an argument for improved management or the use of improved techniques. Improvements in management were seen as a substitute for the lack of modern inputs as a way of raising productivity. A third position agreed that improved techniques were the only basis for long-term progress, but believed that better management was needed to introduce them. Thus, there was not merely a simple opposition of organisation vs. technical change, or relations vs. forces of production.

The state of the technical base of agriculture was extensively reviewed during 1973 and it was criticised for its low quality and inefficient utilisation. The low capacity utilisation of the water control network was reiterated.

In general, only about 60-70% of the prescribed capacity of central projects has been used....Many projects which were built after the restoration of peace in 1954 are now damaged and their quality is generally poor.[108]

The importance of using all fertiliser sources was emphasised as there had been a tendency to rely excessively on chemical fertilisers. Although these were recognised as making an essential contribution to higher yields, they needed to be combined with organic fertilisers for optimum results. One obvious danger of excessive reliance on them was the shortage and irregularity of their supply as compared with the considerable range of organic fertilisers available. US bombing destroyed the major chemical fertiliser plant at Viet Tri in 1967, and subsequently chemical fertiliser was mainly imported.[109] The relationship between animal husbandry and fertiliser availability was emphasised along with the importance of developing specialised fertiliser production teams and expanding the trade in fertiliser.[110] Specialised teams and cooperative pig herds were features of the proposed NMS.

Animal and human excrement provided the major sources of fertiliser and specific collection quotas for them had existed since the early 1960s. Each person over fifteen years of age was expected to collect 50kg of excrement and 600kg of urine per year, while water buffalo, cows, and pigs were expected to provide, respectively, 3.5 tons, 3 tons, and 2.5 tons.[111] Fertiliser collection had long been an important occupation in rural Vietnam though collectivisation had reduced the incentive factor involved. The above quotas were coupled with financial incentives and the award of honours. The production of 'green fertilisers' continued to be encouraged.[112] The oft-quoted targets of 8-10 tons of organic fertiliser per hectare remained barely half fulfilled at 5 tons for rice and 3 tons for other crops.[113]

The third major technical input was seeds. Whereas a number of improvements in seed development had occurred, with the exception of the spring rice strains, these had not been sufficiently generalised. As with fertiliser, the NMS proposed specialised teams to produce an adequate supply of high-quality seeds. Immediate tasks were to develop high-yielding varieties for non-crops[114] and for the tenth-month crop, where the yield was now below that of the fifth-month/spring crop.

The fact that detailed agricultural research occurred,[115] but was either implemented selectively, inadequately, or not at all, indicates a gulf between principle and practice not always appreciated by observers, both in the case of the DRV and elsewhere.[116] Possible explanations are that success in certain areas was countered by failure and actual retardation in others; that the extension of the cultivated area to more marginal lands countered yield improvements on the existing area (although the cultivated area actually fell from the mid-1960s); that the provision of extension services was inadequate, partly because of the cadre shortage; that the technical developments on special plots were inappropriate and difficult to generalise; and finally, that adoption incentives were inadequate.

As regards agricultural mechanisation, there were reports of an ever-increasing number of cooperatives with 'mechanised stations,' though how advanced these were is not always clear. Of some 11,000 or more (over half of all cooperatives) in the early 1970s, only some 1,100 had been provided with electricity. The degree of effective utilisation was low. Tractors were being used to only 42% of their capacity. This was attributed both to labour problems and technical deficiencies.

> In particular, the mechanical equipment designed to serve animal husbandry and to process agricultural products was still inadequate.[117]

The mechanisation and semi-mechanisation targets at this time were more realistic than those of the early 1960s. They were to increase support for the water control network; to improve the quality, utilisation, and repair of existing equipment; and to raise the technical level of cadres in various branches. It was irrigation that attracted the most attention and the major share of agricultural investment (see Table 5.3).[118]

The demand placed on labour by the war may have given an impetus to increased mechanisation and the introduction of labour-saving techniques. The seasonality of labour use in agriculture meant that despite a general rural surplus there could be particular labour shortages. At the same time, the effect of the war on industry increased the scarcity of manufactured inputs. The reduction of the official agricultural workforce was particularly dramatic between 1965 and 1968 when it fell by more than 1 million--a little under 18%. As a proportion of the rural workforce it fell from 87.8% in 1960 to 82.1% in 1965 and 62.1% in 1975. Although employment increases occurred in construction and non-material production (mainly education and health),[119] there is still a substantial increase in the percentage of the rural workforce not accounted for. This may reflect war-related tasks or increases in unregistered employment.

The demands of the war on male labour put increasing pressure on the female workforce in agriculture, already over 50% of the total. The need for greater utilisation of female labour brought a greater awareness of and attention to the reproductive tasks carried out by female labour.

> Women are the principal production force at present, but favourable conditions such as solutions for the problems of child care, meals, refreshment, marketing, etc., must be provided so as to mobilise all women's forces to participate in production.[120]

The reduction in the agricultural, though not in the rural workforce, exceeded the output fall and showed an increase in agricultural productivity.[121]

Post-war rehabilitation and a return to 1965 levels in most sectors was the main objective for the latter half of 1973.

During the last six months of the year, a task of important significance is to inspect the economy in all respects. We must round up all equipment, materials and fixed assets: reorganise the technical forces consisting of workers and cadres; reassemble the social labour forces, including professional labourers; etc. By satisfactorily conducting this inspection, we will have a stable foundation for devising positive measures to mobilise the above forces in the restoration and development of the economy.[122]

Judging by the situation in the construction industry, the inventory of State property made in October was badly needed.

During the past 9 months a construction worker worked on average only 19.7 days a month. Days when workers did not work or were absent from work without good cause were greater than before 1971. Labour output was low. Supplies received or issued at worksites were not gauged, measured or counted. There were no planned norms for the use of supplies, or supplies were lost, damaged or pilfered. The number of machines operating at the worksites accounted for only 22% of available machinery. Most machines operated only one shift a day: 21.9% of machines were not functioning: machine and shift output was low.[123]

The years 1974 and 1975 were envisaged as the restoration period from which the economy could go forward.

By the end of 1975 we must produce most of the people's consumer goods, gradually increase the source of capital accumulated in the country and the source of commodities for export in order to establish the conditions for ensuring expanded reproduction and developing the national economy in a balanced, positive and steady manner.[124]

During this period there was an attempt to shift away from policies of self-sufficiency towards greater specialisation.[125] In agriculture this partly reflected concern over the falling production of subsidiary crops and vegetables which affected the availability of human and animal foodstuffs.[126] There was also renewed emphasis on the relocation of the population from the Delta to develop the highlands. This had been notably unsuccessful in previous years.

> During the past decade the cultivated area of the mountain region has not increased and the cultivated area per head has decreased.[127]
>
> The value of products sold to the State has only represented 14% of the total value of the goods produced by the mountainous areas.[128]

One new development was the establishment of twelve 'New Economic Areas' in 1973.[129]

Agricultural production in 1973 was below that of 1972 due largely to extensive flooding of the tenth-month crop. This again emphasised the deficiencies in the water control network.[130] Although similar to those of pre-1965, the economic problems facing the DRV had been worsened by the war. At a macro level, the crisis of overaccumulation and open inflation, was summed up as follows:

> At present our economy lacks a balance between the requirements of production relations and production forces, between the accumulated capital and expenses, between the workforce and labour productivity, between exports and imports, between the requirements for construction and material supplies, and between supply and demand.[131]

The balance between accumulation and consumption was recognised as 'increasingly acute.'[132] The crisis was attributed to three factors: the underdeveloped material and technical base, the war and natural calamities, and the inadequate management and organisation.

The reconciliation of these imbalances and the role of agriculture were again debated in 1974. The issues involved were virtually identical to those of the 1968-69 discussions and to those of 1958-59. The argument continued that many problems resulted from the only partial implementation of the 1969 Statute. One of these problems was inevitably the family economy. It had misappropriated cooperative land and had work contracted to it by cooperatives/brigades. Another was brigade autonomy and pursuit of interests not necessarily in accord with those of the State.

A speech by Truong Chinh, in which he made reference to the forthcoming Statute and Land Law, is quoted at length as a clear description of the cooperatives and attitudes towards them.

> Many agricultural production cooperatives have failed to comply with the Cooperative Statute, have assigned work to households on a contractural basis, have filed inaccurate efficiency returns, have failed to carry out distribution on the basis of the amount of labour contributed, have failed to share benefits in an equitable and rational manner, and have given rise to corruption and waste. The cooperatives' material and technical bases remain too weak. The accumulated funds - or individual capital - remain insubstantial. Many cooperatives have improperly used this individual capital and State loans for non-production purposes. Our State has not yet promulgated a land law. Therefore, a situation has arisen in which land has been used on a scattered basis, and cooperative and State land has been encroached upon. A number of specific policies on the purchase of agricultural products and prices have not yet been modified or rationally supplemented. A number of good cadres experienced in leading the cooperatives and production have assumed other jobs, thuus weakening the cooperatives

> leadership body. We have slackened to some extent the regular education of peasants in the need to wage a thoroughgoing struggle between the two paths. Party branches are weak and their performance is unsatisfactory. They have failed firmly to grasp Party lines and policies and are not exemplary. Generally speaking, we have failed expeditiously to review and disseminate the experience gained by the cooperatives and localities which have satisfactorily carried out their tasks. We have also failed expeditiously to analyse and criticise the unsatisfactory performance of various localities, so as to correct their shortcomings.[133]

Other articles cite cooperatives where up to 30% of the land had been appropriated by cadres and members.[134] 'In many localities cooperatives are declining, some of them have collapsed or are only carrying out activities for form's sake.'[135]

The new Statute was adopted by the Politburo in March 1974 and published in October. It again stressed <u>collective</u> control over the utilisation of resources in order to raise both labour productivity and labour payments and hence increase the incentive for peasants to devote labour time to the brigade. This created a vicious circle because in order to derive the alleged scale advantages, all peasants had to participate fully in collective activities. In addition, collective control would satisfy the immediate objective of raising output. This might, however, encounter diminishing marginal productivity of labour time and conflict with the first objective.

> Our manpower is still abundant in the countryside. The most urgent need for our country is quickly to increase our gross output, in particular food output. That is why in many cases we have fallen into a situation in which our efforts have far exceeded the benefits gained.[136]

Contemporary reports continually emphasised

the limited mobilisation of labour by cooperatives, the low labour productivity, and hence low value of workpoints. This failure to raise income was the crucial constraint on raising peasant participation.

> In many areas, the labour of cooperative members has not yet been fully utilised in production, and in busy seasons, even comparatively outstanding cooperatives have only been able to mobilise about 80% of their available labour force. This ratio has dropped to 25-30% in less busy months. On average, each worker has worked at most about 200 days a year for the cooperative. The number of actual working hours is small, being about 5-6 hours a day; and the number of hours in which the work is performed with good quality and in accordance with proper techniques is even smaller.[137]

This report describes practices of dishonestly claiming workpoints and workdays which further reduced the value of a workday. The variation in payments is considerable--0.2 to 1.0 *dong*--with the highest revealing very little increase over the levels of the early 1960s.[138] Those cooperatives with high workday payments may have achieved these by maximising labour productivity rather than output. Those receiving these payments might have been opposed to increases in the cooperative labour force which might dilute their income.

The overall conclusion of these reports is that no significant improvement in labour mobilisation and utilisation in the collective sector had occurred since the earliest days of its development. Nevertheless, efforts were continued to convince the peasantry of the permanence of this form of organisation, of its material superiority over family production, and that increased application of more effective labour would, at least in the medium term as technical improvements were introduced, produce rewards commensurate with effort.

Criticisms of labour productivity, misuse of

resources, and of insuffieciently 'socialist' attitudes to work were not confined to cooperative agriculture. They were made also with respect to State farms where productivity was apparently below pre-war levels,[139] and to other branches of the economy. The Statistical Branch suffered particular criticisms. It received inaccurate primary data for a variety of reasons.

> Statistical data concerning rice yield and production volume are exaggerated to magnify achievements, but they are trimmed down when the time comes to fulfill the grain obligations.[140]

This again indicates the conflict between State and peasantry and the ability of the cooperative or brigade to represent the peasants' interests. The following provides an overall description:

> In many units, there are more personnel, more machines and material supplies and more State investment than in the prewar period; yet there has not been a production increase, labour output has decreased and there have been losses in production. Generally speaking, due to the low labour productivity in our society, our labour still fails to provide us with sufficient food to eat, and we still fail to achieve capital accumulation with which to expand our reproduction. Because our labour still fails to meet our consumption requirements, our production and living standards are still facing difficulties.[141]

The economy was unable to break out of a vicious circle of low productivity, hence low surplus, low accumulation, limited technical progress and ability to provide production incentives, and hence continuing low productivity. The population census of 1974 revealed an increase of almost 50% since 1960--an annual increase of 2.9%.[142] To provide an adequate living-standard and possibly induce a lower population growth rate required increases in production and productivity, especially in agriculture. The State had believed this could be achieved by curtailing consumption and by an emphasis on industrial accumulation. This had failed to produce the rapid technical progress necessary to develop the dynamics of this growth process.

> If production fails to bring about an accumulation, there will be no expanded reproduction and it will be impossible to step up socialist industrialisation. For this reason production must be connected with economy measures and a display of economic efficiency.[143]

Foreign aid was not sufficient to significantly reduce the squeeze on consumption. 'Our fraternal countries' assistance is important, but cannot replace our people's own efforts.'[144]

The centrality of cooperatives in this process derived from their alleged ability to achieve production and productivity improvements, to facilitate procurement and hence accumulation, to utilise underemployed labour, and to limit consumption. Justification for these arguments remained elusive but was believed to lie in increased organisation and control of the peasantry.

> Experience gained from many campaigns indicates that if the cadres <u>assigned</u> to a cooperative <u>take over</u> all of its affairs improvement will be noted, but after they leave, everything will return to the former situation. The cooperative will remain stagnant.[145]

<u>The Thai Binh Conference</u>

A Conference on the Reorganisation of Agricultural Production was held in Thai Binh province from August 5-12, 1974. This province had consistently achieved the best agricultural performance and was the first to breach the five tons of paddy per hectare per year barrier. The publication of the important Conference speeches was accompanied by a VWP CC Directive and the Draft Statute on High-Level Cooperatives. In addition, the Council of Ministers issued a communique on agricultural price policy.[146]

The situation in which the Conference took place was characterised as follows:

> The productivity of many products has not increased, or has increased very slowly; the cultivated area has decreased; agricultural production has failed to meet the people's demand for grain and food, and has failed to provide sufficient raw materials for industry; the volume of agricultural products available for export remains insubstantial.[147]
>
> Production relations in many localities have been disrupted because the income obtained from the collective economy has not increased and, worse still, has decreased, distribution has not yet become equitable and, consequently, many cooperative members have been attracted into family-based economic activities and into the free market; and the management of the cooperatives and the socialist market has encountered more difficulties.[148]

Many of these problems should have been dealt with when the cooperatives were formed, or when they were raised from low- to high-level:

> Thus, we are carrying out these tasks about 10 years later than they should have been carried out.[149]

They would now make a further attempt to devise policies enabling collective agriculture to meet State objectives and realise its alleged potential. The immediate objectives were to solve

> in 3 or 4 years, the people's need for grain and food in their daily life and to improve their diet,

and, to ensure,

> raw materials for the factories that have already been built or will be built in the next few years...and to raise, within the next 5 or 6 years, the value of ex-

> ported agricultural products to the same level as or above the level of imports needed for agriculture.[150]

With regard to labour the objectives were,

> to motivate everyone to work enthusiastically for 10 to 12 hours a day and constantly increase labour productivity; and to help them strive to attain on a widespread basis the following objectives: one labourer should cultivate one hectare, achieve a rice yield of 5-6 tons/ha. of two-crop ricefields,and raise two hogs weighing 50kg or more each at the time they are removed from pig-pens.[151]

This would aid the redistribution of labour within agriculture and between sectors.

The extent to which industry could assist fulfillment of these objectives was extremely limited.

> According to a preliminary assessment of the situation prevailing in various sectors, our industrial capabilities, especially those of the machinery branch, will remain limited for the next 3 or 4 years, and the possibility of importing equipment and machines for our agriculture will also be limited. Therefore, in developing our agriculture we will continue to rely on manual labour and will primarily use improved tools.[152]

Development of water control, fertiliser, and seeds was to be based on a more generalised dissemination of existing best-practice techniques. This would be related to changes in management and organisation which, it was believed, could lead to a greater realisation of the latent potential of agriculture.[153] 'Improved' management was seen in part as more important than technical development, but also as a basis for the latter. Realistically, given the constraints on the supply of inputs and incentives, it was the only possible solution within the framework of the overall development strategy.

However, given the range of policies which had been applied in previous years and the continual management 'improvement' campaigns, there was little new to consider. Hence, either the campaigns and policies had been ineffectively implemented or the underlying premise--that there existed a considerable latent potentiality within an agrarian sector based on collective production--was incorrect. Since admission of the latter was inconceivable, 'new' policies had to be predicated on the first assumption and consequently they appear as previously-tried policies in new garb.[154]

Specific price adjustments occurred which included raising the purchase price of livestock to encourage meat production; raising the purchase prices for some industrial crops because of their higher production costs and export significance; and lowering the prices of a number of inputs. Differential regional prices were introduced for grain and subsidiary crops to help areas encountering difficulties and to encourage supply for the food-processing industry, but basic grain purchase prices remained unchanged.[155]

In a passage apparently added to answer criticism by delegates, Le Duan made clear that there would be no major change in pricing policy:

> Some of you have told the Conference that <u>prices constitute one of the reasons cooperative members have displayed a lack of enthusiasm in production</u>. It may be correct that the purchase prices of some agricultural products are still unreasonable and the State must resolutely adjust them. However, we must take into account the fact that peasants purchase industrial products from the State at relatively low prices. Generally speaking, if we increase the prices of agricultural products, we must also increase the prices of industrial products, and realities have shown that the quantity of products purchased by the State from peasants is smaller than the quantity of products purchased by the peasants from the State.[156]

State.[156]

According to official statistics, the State accounted for just under half of all peasant purchases and only 30% of peasant sales.[157] This behaviour was understandable if State prices for both purchases <u>and</u> sales were lower than those of the free market. The price of agricultural production materials sold by the State rose 8% between 1964 and 1973 compared to 68% on the free market. Food prices rose 17% and 44% respectively.[158] Peasants would sell on the free market, which had doubled its share of trade since 1968,[159] and use the cash to buy from the State.

Le Duan draws the wrong implication, for changing the State terms of trade in favour of agriculture might have increased peasants' sales for the State both absolutely and relatively. His reference to raising industrial prices in line with agricultural prices fails to take into account differing price elasticities of demand and differing productivities in the two sectors. In a market economy, assuming constant returns production functions and with more rapid technical progress in industry, agricultural prices would rise <u>faster</u> than industrial prices to equalise increases in value productivity.

Increased sales were to be encouraged not by higher above-quota prices, but by action against the free market in grain and by the signing of bilateral contracts with cooperatives and individual peasant families.[160]

The Draft Cooperative Statute was based on the 1969 Statute and subsequent experience. It attempted to deal with some of the 'weaknesses' referred to by Truong Chinh and by speakers at the Conference. It provided strict stipulations on the establishment of production brigades and teams and the number of workdays to be contributed by them. It permitted the subdivision of production teams into small labour groups for particular tasks, but prohibited the allocation of work to individual families [Article 20(4)]. It strictly stipulated the family-economy relationship with the cooperative in terms of the land allocated, its location, the uses to which it could be put, and obligations to be fulfilled by the family with respect to fertiliser (Article 19). The importance of the family-economy for animal husbandry was emphasised by Le Duan who stressed that 'it can make full use of under-utilised labour and food remnants in each family.'[161] The encouragement of collective livestock herds was, however, a feature of the NMS.

One alleged weakness of cooperatives was their size. This was to be raised to 200 hectares and 500 working members on average. Cooperatives currently averaged just under 200 families and 300 working members. The major criticism, however, concerned the problem of control--State/cooperative, cooperative/brigade, and brigade/family. The search continued for a level or form of organisation which would overcome peasant independence and serve as a unit of _effective_ control over production, distribution, and consumption.

> Though being socialist economic units, our cooperatives still manifest the characteristics of guilds. Like an independent organisation, each cooperative rarely entertains relations with another cooperative or even with the national economy. Each production team in the cooperative also bears the same independent nature.[162]

The New Management System (NMS)[163] was based on a unification of economic activity, management, and distribution in the hands of the Cooperative Management Committee. All aspects of planning, production, resource utilisation, distribution, and allocation of consumer goods were to be controlled by this central authority and _not_ by the brigade. In policy statements, especially at the Thai Binh conference, the district was to take over the 'unifying' role by establishing agricultural committees 'to exert effective, uniform control over agricultural activities and agricultural cooperatives.'[164]

> It is obviously impossible to reorganise production within the framework of a single cooperative having a few hundred hectares of land and a few hundred work-

> ing members. This can be achieved only in an area where there are a large number of workers and more land. I think that under the present conditions and with its available equipment and machines, a locality or district having about 10,000 hectares of ricefields and 40,000 agricultural workers, will be able to redistribute its manpower and reorganise its production better and more rationally.[165]

The Cooperative Management Committee was to prepare an overall plan to be approved at the district level and incorporated into a National Plan. It thus presupposed the existence of effective cooperative level management which, based on this plan, allocated resources to brigades and teams by way of various forms of contracts.

Brigades and teams were to be specialised by task(s) rather than the standard existing form of general brigades. Specialisation was intended to yield economies *and* curtail brigade autonomy-- the cooperative management would be more subject to State control (there were only 18,000 cooperatives by 1974). By establishing brigades for seed development, livestock, and non-agricultural activities, more functions would be incorporated within the cooperative and its 'plan.' The 'triple contract' was retained, but the number of work norms and extent of supervision were to be substantially increased and brigades were to be totally dependent on the cooperative for all inputs. As frequently emphasised, distribution was to be based largely on workpoints rather than on an *ad hoc* basis as in many cooperatives.

The NMS was introduced in experimental 'model' cooperatives in the early 1970s and, in these cases, largely achieved the intended objectives. The autonomy of the general rice-producing brigades was curtailed in that production was increasingly controlled and planned, and substantial increases in output and marketed surplus were achieved. However, the 'models' tended to have favourable resource endowments and were based on cooperatives with a good track record of central management. Following the Thai Binh Conference, which had assessed the current state of cooperative development and concluded that many had, at best, a nominal existence, the NMS was to be widely introduced.

The main reasons for its subsequent failure (or non-implementation) should be clear without detailed analysis. There was an extreme shortage of State-supplied inputs. The resource base of the majority of cooperatives was inferior to that of the models. Efficient and 'honest' managerial and technical cadres were in short supply. The very factors that determined the limited functioning of existing cooperatives, such as lack of incentive and brigade and family independence, militated against the functioning of any 'improved' system, and finally, by the mid-1970s, there was a flourishing free market, to which peasants would be unwilling to sacrifice access.

1974 was an extremely good year for agricultural production with paddy output surpassing even that of 1959, although per capita production remained substantially lower. Despite arguing that no straightforward solutions existed, the leadership continued to propose the same policies with only slight modifications, on the grounds that they had not been effectively implemented. The solutions to the problems posed by low output and marketable surplus and independence remained larger units under greater State control and 'improved' management. The NMS was the last in a series of attempts to demonstrate the superiority of cooperative over independent family production in existing conditions in the DRV, and to find an effective means of 'taxing' agriculture.

The essential dilemma of developing agriculture along socialist lines in an underdeveloped technical environment remained and was only partially confronted. Le Duan asked whether there existed a special law for the development of agriculture in such a situation, whether such a law were specific to the DRV, and whether its nature had been fully grasped.[166]

A somewhat different picture of reality than the one apparently held by the leadership is provided by the following comment, which is actually discussing VWP cadres.

> As inherently small producers who are still engaged in small-scale production, all of our comrades are still heavily influenced by the thoughts and habits of small producers and small owners. In many cases a considerable number of our comrades, in assessing things, usually adopt the standpoint of small peasant producers and are normally concerned with the personal, selfish interests of small proprietors. The vestiges of feudalism have not yet been completely eradicated.[167]

The gap between the form of 'socialist' cooperatives and the reality of small-scale agricultural production--a significant element of which took place at the level of the family--remained as wide as ever.

Summary

This chapter has covered a longer time span than the preceding chapters and includes the period of war with the United Sates. One major consequence of the war was a relocation of industrial capacity but with retention of the emphasis on heavy industry. Not only did light industrial production continue to grow much more slowly than heavy industry, but the neglect of handicrafts continued.[168] Another consequence of the war was increased inflation and a further fall in living-standards.

The war exacerbated the existing problems of agriculture. The leadership continued to believe that it was possible to raise agricultural output with traditional inputs within cooperatives, despite the lack of evidence to support this belief. The virtues of cooperatives continued to be emphasised, with their shortcomings blamed on inadequate leadership and management, and the inherent backwardness of the peasantry in technique and ideology. The urban-based VWP leadership appeared to hold a disparaging view of the peasantry.[169]

The relative failure of cooperatives meant

continued State dependence on household production, which was also shown to be important to the peasant families. This dependence meant a continuing use of material incentives and at least tolerance of private production. The resulting dilemma was similar to that faced in China:

> how to treat the private sector in such a way as to secure the maximum cooperation from the peasants and guarantee the economic progress required in collective agriculture.[170]

The increased degree of autonomy resulting from the war difficulties had actually reduced State influence and control and led to increasing *de facto* privatisation of production.

Throughout the chapter reference was again made to household decision-making and the problems this posed for State objectives. Although data are limited, it seems clear that average productivity on cooperative land was lower than on household plots, partly because of supervision difficulties, and that marginal productivity was below average productivity--with respect to both land and labour--with existing inputs and technology.[171]

The chapter shows that there were no new policies during this period. The same ones were reapplied and reemphasised with an underlying belief that some mix of policies could succeed in raising agricultural production and removing the constraint on industrial growth.

One positive development was the development of 'spring' rice, although even this had some negative impacts on other crops.

One aspect of the leadership's beliefs in socialist forms, seen in this and the previous chapter, was that if relative superiority had not yet been demonstrated it would be--by bigger, better-managed cooperatives. This explains the move to high-level cooperatives, new statutes, management campaigns, and later focus on the district. These moves were, of course, aimed at validating the arguments for the productive superiority of cooperatives and ensuring State control over production and surplus.

The chapter ended with discussion of the Thai Binh conference in 1974. There was a greater awareness of the problems of the economy and agriculture, but a repetition of the same arguments in favour of cooperatives which had been used for fifteen years. It seemed that there was a choice between accepting that the framework of agricultural policy was inappropriate, which was inconceivable given the political system, or arguing that the policies were right but had been inadequately implemented. This 'failure' was due to the deficiencies of the implementors and the peasants themselves. An example was the pricing system which was recognised as causing problems, but which the leadership at this stage refused to rationalise. The following comment from 1982 would have been inconceivable in 1974.

> There were prices, but they were artificial. They did not represent the real value of the thing concerned....From now on, the relations between the State and our peasants will be governed by real prices.[172]

The Thai Binh conference discussions provided further evidence of the reality of agricultural production and organisation and its divergence from the form and the policy.

Conclusion

Summary

This work has reviewed the history of agriculture in the DRV throughout its existence as an independent state. It has focussed attention on the institutional organisation of agriculture while also detailing its actual performance. Policy has been assessed largely in the light of this performance.

At the beginning of the period agriculture was suffering from years of neglect and the ravages of war. It was characterised by very small holdings, extreme inequality, and a degree of landlessness.

The agrarian policy adopted by the VWP followed communist orthodoxy in that agriculture was of secondary importance to industry and was to be organised along collective lines. It followed the Chinese variant inasmuch as mechanisation could follow collectivisation. The ultimate objective has always been, and remains, a fully collectivised, mechanised agriculture.

In the early years of its existence, the DRV was fighting against the French for independence. The primacy of the military struggle meant that little attention was paid to implementing agrarian policy, and also dictated the nature of that policy. Thus, even a radical distributionalist land reform was not an immediate objective because of the danger of alienating surplus producers, the limited geographical area under permanent DRV control, and the shortage of suitably trained cadres. However, prior to 1953, some reforms of taxation and rents did take place as well as the distribution of some land.

Once the situation for both the country and the ruling party was more secure, the move to a distributionalist land reform was made. This broke completely the economic and political power of the landlords, honoured the promise to the peasants, and was a necessary first step in the institutional transformation of agriculture. The Land Reform produced a transformation of agricultural holdings but encountered a number of problems. These resulted ultimately from the absolute land scarcity in the DRV. This meant that there was not enough land to provide subsistence holdings for all rural households without expropriating a large number of families with slightly above-average holdings. Reliance on poor peasant cadres as the 'activists' of Land Reform produced the pressure to do just this, while the ambiguity of the classification criteria, the continued socio-economic mobility of the peasantry, and the temporary distribution of land in previous years made conflict inevitable.

It was shown that conflicts within the peasantry resulted in political and economic problems, and the severity of these conflicts and their potential implications made some corrective action imperative. The timing of this action was influenced by developments in the wider communist world in 1956. The corrective action stabilised the situation in agriculture and the VWP, and there followed an expansion of production on the basis of small-holder agriculture in the second half of the 1950s.

It was argued on the basis of available evidence, that a 'scissors crisis' emerged as agricultural production grew faster than the availability of consumer goods. This led to a lower level of marketing of agricultural produce than was desired by the State, which illustrated the unacceptable--to the State, VWP, and industrial strategy--independence of the peasantry.

The need to control agricultural marketing and surplus, to make agriculture subordinate to industry and, crucially, to prevent it from acting as a constraint on industrial expansion, was one justification for collectivisation. This was clearly shown using quotations from party documents of the period.

The other main arguments for collectivisation of agricultural production were that it would limit differentiation and the transformation of small-holder peasant agriculture into capitalist agriculture, and that only collective production could produce rapid growth by taking advantage of the larger scale of production involved. This was alleged to be true for both existing technology and new technology. Not only would cooperatives be more appropriate to mechanised agriculture (the orthodox Lenin/Stalin argument), but they could also more fully exploit existing resources, essentially through the larger scale of organisation and a greater division of labour (the Chinese amendment to orthodoxy).[1]

There was thus both a static and dynamic argument underlying the arguments for collectivisation. It would enable the State to gain a larger share of a given amount of agricultural production and also enable the total output to grow faster than otherwise. Clearly, if these arguments were valid, then growth could have been rapid and unconstrained by agriculture. In detailing agricultural development and performance from 1960 onwards, attention was paid to the extent to which these arguments were valid.

Again, on the basis of limited evidence, the differentiation argument was criticised. The 'supporting' evidence cited by the VWP in 1959 does not show a pattern of increased differentiation, and later evidence does not show that collectivisation clearly reduced differentiation. It was asserted that peasant socio-economic mobility was the result of a range of factors of which the post-Land Reform holdings was only one.

The dynamic argument that only cooperatives could realise the current and future potential of agriculture was not substantiated by the subsequent fifteen years of agricultural history. Various reasons offered by the DRV leadership as to why this argument was not validated were examined in the course of presenting the history. This showed that the leadership never seriously doubted the ultimate validity of the argument and saw the problem as one of inadequate policy implementation or the incorrect combination of policies. The framework never appeared to be in serious dispute. Such disagreements as existed within the leadership were over the relative efficacy of different policies and the timing of certain stages of the institutional transformation.

Reasons were considered for the invalidity of the argument for the superiority of cooperative as opposed to peasant household agriculture. The nature of peasant decision-making, the non-existence of economies of scale, the existence of scale diseconomies, and the relative primacy and productivity growth potential of agriculture and industry were all areas of attention. Empirical 'evidence' on these aspects is clearly limited; firm substantiation would involve greater use of comparative material given the shortages of DRV data. However, as the postscript to this chapter will show, subsequent Vietnamese analysis and experience accord with the interpretation given here.

It was argued that the actual failure of collective production to perform as the leadership hoped, created further complications. It made it more imperative for the State to increase its control over production and surplus in order to extract an increasing amount. It thus strengthened the control argument for collectivisation. At the same time, it increased State dependence on non-collective production to make up the shortfall. This required at least tolerance of private production. This very tolerance and the ability of peasants to earn proportionately more from this activity reduced the incentive to participate fully and enthusiastically in cooperatives, and hence reduced any slim chance of them demonstrating superiority. This same point has been made with respect to the USSR.

> The benefits of the high free market prices accrued to the private sector of agriculture, i.e. peasants who had not yet been collectivised, regions where agriculture had not yet been fully collectivised,...and those collective farmers who had a marketable surplus from the output of the private plots. Naturally, this helped to weaken the collective farms.[2]

The incentive to maximise output was also constrained by the limited quantity and quality of consumer goods available.

Evidence on this range of issues--State procurement, prices, availability and quality of industrial goods, dependence on private production, the extent of private production, the relative productivities of private and collective production, and peasant attitudes--was not of the hard empirical kind. In most cases more data would be required to fully substantiate the contentions made. However, it is believed that all the non-empirical evidence of many different kinds supports the contentions here, and that further data are likely only to strengthen the argument. One obvious reason for this is that reliance has been placed completely on 'official' material which, it must be assumed, most closely supported the 'official' argument.

The virutal stagnation of agriculture in the DRV during the period 1960-75 is here attributed to the very policies pursued and to the framework in which they were embodied--heavy industrialisation and collective agriculture. It could be argued that the leadership was wrong, not as regards the framework (i.e. collectivisation), but in the belief that there was scope for increased output and productivity with traditional inputs. In the case of China it has been suggested that this scope *is* limited and that the sources of such growth that occurred were modern current inputs, at least after about 1957.[3] Evidence from other countries would also support such a view.[4]

Subsequent experience in Vietnam and China suggests that both arguments may be valid. There is still scope for expansion with traditional inputs, but within a framework of family production and exploitation; and also that greater attention to agricultural investment, research, and inputs produces results.

The final section showed how the Vietnamese themselves began to consider such a viewpoint and to reconsider certain basic elements of the policy. Throughout the period up to 1975, they attributed the virtual stagnation to inherent backwardness in all senses and to a failure to implement the prescribed policies adequately. After 1979 this attitude altered as 'more appropriate' policies were devised and some criticism was made of the earlier policies.

Postscript: A Vietnamese Reappraisal

In the late 1970s, the Vietnamese themselves made major criticisms of the system of agricultural organisation in what was, by then, the northern part of the Socialist Republic of Vietnam (SRV). Some had held a critical view earlier:

> When changing from hamlet to village cooperatives, if production does not develop, the material and technical base does not develop, the division of labour does not develop, and the labour productivity does not increase, then we cannot regard this new form as suitable. Because it is not suitable and does not satisfy the development requirements of the productive forces...temporarily reverting to lower forms is more favourable for production and socialism.[5]

As has been shown, these developments had not occurred in the majority of cooperatives in the DRV at any of the various levels of development. The NMS and focus on the district was the last in a series of attempts to find an appropriate and effective level of organisation and control. The failure of this last attempt brought forth a much wider debate which questioned not only the returns to scale argument, but also the procurement and 'seeds of capitalism' arguments. The latter was restated by Le Duan in 1974: 'Small production begets capitalism daily and hourly.'[6]

The following quotation from 1979 provides a striking contrast to innumerable earlier comments:

> The simple mentality of wanting to immediately control all production and distribution by administrative laws and regulations and eliminating everything else, not allowing anyone to do anything outside of nationalised industries and cooperatives, forbidding all forms of exchange, can only lead to an economic situation of poverty and slow growth. A number of weak spirited comrades are

> obsessed with the spectre of the spontaneous development of capitalism when any production or exchange is not organised. Petty individual production or petty trade can only develop into capitalism through a process of accumulation of capital and its use in capitalist enterprises, creating private ownership of large-scale means of production. The proletarian State and socialist economy do not allow this to happen.[7]

By 1979 conditions in the north of the SRV had changed considerably.[8] The war with the United States was over and the country had been politically reunified. The 'promise' of substantial US aid had not materialised. The south suffered from the cut-off of US aid, particularly important in consumer goods and agricultural inputs such as fertilisers and pesticides. Relations with China had deteriorated to the point of war, thus cutting off all aid from that source. Attempts to attract private capital had also failed. Severe economic problems had emerged for a number of reasons. They included a series of very bad harvests, the chronic shortage of consumer goods exacerbated by the excess demand of the south spilling over to the north, and the deterioration of the government's financial position and increased deficit financing. A consequence of the economic crisis was inflation. Trade links had shifted to almost total dependence on the Comecon countries which Vietnam had joined, when, as in the late 1950s, Soviet aid had switched from grants to loans on harder terms.

> And all the aid has to be paid for. For example, when they put capital into rubber production, we may have to export rubber for many years to pay for it.[9]

These factors acted as catalysts for changing agricultural policy. However, the shift stemmed from the problems of agriculture in the DRV over the previous twenty years, and involved recognition of many of the fundamental criticisms which have been raised in this volume.

The precise timing of the 'reforms' was determined by several factors. They included the poor harvests resulting from extreme climatic conditions,[10] the economic problems stemming from unification, the need to increase agricultural exports to pay for imports and borrowings, and the increasing pressure of 'reformist' planners as the old guard retired.

Signs of a shift can be traced to the 4th VWP Congress in 1976,[11] although officially the focus remained the NMS and the district. Changes in agricultural policy emerged clearly from the 6th Plenum of the 4th CPV CC in the summer of 1979, while the 9th Plenum generalised many of the reforms, both throughout agriculture and the rest of the economy. The emphasis was on raising production by linking the direct producers with the results of their work, and allowing greater freedom in disposing of surplus production. Prices were increased to provide greater incentives to sell surplus to the State and preferential access to State-produced industrial goods was provided.

In agriculture, the role of cooperatives was changed in principle to one of overall planning, supervision, and support of production, with a greatly modified system of contracting work and resources. The system, as laid down in the low- and high-level cooperative Statutes,[12] had been based on contracting output quotas to brigades and on short-term task contracting. The NMS had involved specialised contracting to brigades. The 1979 reforms led to contracting of output quotas to individuals and families and, more radically, the contracting of land to individuals and families.

In principle the cooperative retained control of '5 tasks'--soil preparation, seed-related tasks, fertiliser preparation, water preparation, and pest control--while the three most labour-intensive tasks--transplanting, weeding, and harvesting were contracted to groups of workers and individual workers.[13] The '5 tasks' could also be contracted out. In certain areas used as 'models' --notably around Haiphong--where mechanisation and collective fertiliser were more developed, the '5 tasks' were still done by production brigades. In other areas, where cooperatives had always been dependent on family-owned draught animals and manure from family pigs, all tasks were contracted out. In practice, product-contracts were introduced with families rather than with groups or individuals.

The cooperatives were encouraged to contract out land not directly utilised or left fallow between harvests. In practice they tended to contract out much more of the cooperative's land. The 'triple contract' system was retained, but with greater flexibility over disposal of the surplus. A uniform penalty/reward rate of 100% of the shortfall or surplus of products turned out by labourers under contract was introduced. For production brigades the rate was both less and graduated.

The subcontracting of land took a number of forms,[14] one of which was land to individual members working alone--either single-parent families, the old, or widowed. A second was land contracted to peasant families with several cooperative members or several engaged in agriculture (these were not synonymous since not all peasant family members necessarily joined the cooperative). It has been claimed that the new system _increased_ cooperative membership and encouraged all members, including cadres, to work.[15] A third was land contracted to families with some members working in agriculture and some engaged in other occupations, but available to assist with particular tasks. The amount of land and work contracted out was dependent on the labour capacity of families or individuals and not on their consumption needs. Land users were prohibited from renting out land, mortgaging, or selling it in any way.[16] Hence, no market in land developed. Peasants were, however, encouraged to buy their own means of production including draught animals.

To a large extent these 'reforms' legalised what had already occurred, although this in itself would have had an impact. It was _officially_ acknowledged that the product-contract system had begun to be applied spontaneously in 1977.[17] Its beneficial effect on output, land and labour use, and productivity was frequently stressed. By the

1980-81 winter-spring crop season, over 70% of cooperatives had apparently applied it.

The introduction of the product-contract system was accompanied by other reforms. Private trade was liberalised and incentive prices were substantially increased--the State purchase price of rice rose eight-fold to 4 _dong_ per kg, although the real increase was lower because of inflation. In addition there was again a shift to local industries but with increased emphasis on consumer goods production. The 6th Plenum of the CPV CC, which endorsed private commerce and material incentives, encouraged family and private production of _all_ goods[18] whose inputs did not compete with inputs in short supply to State factories.

The reforms still represented a compromise between those prepared to make extensive use of the market and individual incentives and considerably reduce the complex and bureaucratic system of controls, and those who believed that the solution was more control and 'improved' management.

Before the official generalising of the product-contract system to rice production, which resulted from the 9th Plenum of the VWP CC in December 1980, Le Thanh Nghi had asserted that if this occurred,

> the mode of collective production will gradually degenerate into a mere exercise in formalism not unlike the three-contracts with families which has been criticised by the Politburo in the past.[19]

The 'reforms' also faced a number of problems, some of which were a continuation of those experienced earlier, while some resulted from the new system.[20] Cooperative cadres themselves increasingly accepted contracts and, while this increased the utilisation of the workforce and reduced 'unproductive' labour, it could also lead to a neglect of management duties. However, only a minority of cooperatives had ever functioned effectively and criticism of the cadres' behaviour under the old system was considerable:

> Cooperative cadres tried to line their

> pockets in one way or another and seized every opportunity to wine and dine out of every fund.[21]

Many of the larger cooperatives were subdivided and production brigades broken up into smaller groups which might reduce direct State control. However, the new system was likely to make the planning system work better by creating an environment in which production could increase to fulfill targets rather than as in the previous system, which was summed up by one cooperative member:

> You've no respect for democracy. You plan everything at your desks and force us to comply, although your plans are not practicable. Take rice 424. It's short-stemmed, but you had it planted in low-lying fields, and we ourselves had to bear the resulting losses.[22]

The new system also had implications for employment:

> By implementing product contracts with individual workers, many cooperatives have utilised their workforce better than before. As a result they have some surplus workforce.[23]

The rural labour surplus would thus appear as open unemployment[24] and emphasised the need for alternative rural employment opportunities.

Unemployment could in turn have consequences for differentiation. Most policy statements of this period stressed the need to retain adequate cooperative welfare funds and to make provisions for those who lacked labour power and were unable to work.[25] This related to a concern for retaining many of the functions of the cooperative even though, with a higher level of production and surplus, cooperatives were more likely to be able to provide rations for those in need. In the past, cooperatives often had great difficulty maintaining the various funds. In general, differentials were recognised as a positive feature:

This difference in income is now a motive force spurring all units and labourers to develop production.[26]

Related to welfare and differentiation, the position and status of women and the possible exploitation of family labour in the new system provided cause for concern.[27] However, to the extent that incomes increased, improvements in conditions occurred. Contracts were with individuals and, in theory at least, provided equal opportunity for male and female labour.

In official presentations, the new system was contrasted with the old based on the fiction that the system actually worked. Given that to a large extent it didn't, comparisons (over employment, differentiation, etc.) can only be made with the *real* previous situation. In practice, the post-1979 reality may also have diverged from the new 'form.'

At the macro level the State shifted from a focus on procurement as the priority to a focus on raising production, believing that this would itself raise procurement. In the short-term, the new incentive prices and relaxation of trade controls had inflationary consequences because of supply shortages of industrial goods from the urban sector. Reduction of these pressures, and continued stimulation of agricultural output remained crucially dependent on increasing the supply of consumer goods and inputs to agricultural production. The State introduced a form of grain-fertiliser barter system[28] in which those who sold to the State and met quotas received preferential access to inputs. The effectiveness of this depended on fertiliser availability which was constrained by domestic production and an extreme foreign exchange shortage.

However, despite these unresolved difficulties, the reforms recognised a number of major problems which had confronted cooperatives.[29] Output per hectare on family land was twice that on the cooperative land. The value of the brigade workday was falling, according to Le Duc Tho, because the old 'triple contract' did not link output to each worker.[30] The number of effective cooperatives was at most only 20-30% of the total. The 'material and technical' base of agriculture was inappropriate to the collective form of production relations and will be for some considerable time. Cooperative management was dogged from the start by inexperience, bureaucracy and 'corruption,' while family production was 'economically efficient.'

> We must now uniformly contract out <u>all</u> crops, animals, trades and occupations in cooperatives so that every sector will attain high economic efficiency.[31]

Increasing quantities of industrial goods had to be made available to encourage agricultural production. The price system had to be rationalised --partly reinstating the law of value. Handicraft production was to be encouraged both to provide goods for the peasants and to provide alternative rural employment opportunities. This suggests a principled shift from blind adherence to the Stalinist heavy industry model.

> Until 1979 we didn't pay enough attention to handicraft production.... No doubt, some of the leaders thought the quicker we got rid of handicrafts, the better...for a long time to come, we must consider handicrafts a main branch of production.[32]

Arguments regarding the 'spirit of private ownership' and the need to prevent this from leading to class division and capitalism were repeated and were a legitimate concern. However, the '3 interests' of peasant, cooperative, and State were explicitly recognised, as was (above all)

> the truth...that the labouring people... are simply not as interested in working for the cooperatives as they are in concentrating their efforts on other work activities in order to increase their income and ensuring their livelihood.[33]

Table I - Change in Landownership and Use, 1945-53

	French	Church	Common	Land-lord	Rich Peas-ant	Middle Peas-ant	Poor Peas-ant	Land-less Peas-ant	Other Rural Work-ers	Total/ Average
1945										
Families	----	3	----	1985[1]	1278	18422	20246	11785	3480	57199
% of all families	----	----	----	3.5	2.2	32.2	35.4	20.6	6.1	100
Land owned (ha.)	346	692	10743	9842	3115	13266	4734	545	328	43611
% of land owned	0.8	1.6	24.6	22.5	7.1	30.4	10.8	1.2	0.7	99.7
Land owned per family (m^2)	----	----	----	49582	24374	7201	2338	462	943	7624
1953										
Families	----	3	----	1311[1]	854	23774	30706	10216	3736	70600[2]
% of all families	----	----	----	1.8	1.2	33.6	43.5	14.4	5.3	99.8
Land owned (ha.)	420	687	9871	7586	1929	15154	6863	936	350	43796[3]
Land owned per family (m^2)	----	----	----	57864	22588	6374	2235	916	937	6204[4]
% of land owned	1.0	1.6	22.5	17.3	4.4	34.6	15.7	2.1	0.8	100
Dispersed French land (ha.)	143	----	----	----	----	57	163	53	3	419
" Common " (ha.)	----	7	630	217	150	3499	4108	1120	140	9871
" Landlord " (ha.)	----	----	----	2	22	1160	1627	571	31	3413[5]
" Rich Peas-ant " (ha.)	----	----	----	----	----	62	59	28	1	150[6]
" Church " (ha.)	----	270	----	2	3	141	201	67	3	687
Total land dispersed (ha.)										
- by recipient	143	277	630	221	175	4919	6158	1839	178	14540
- source	419	687	9871	3413	150	----	----	----	----	14540[7]
% dispersed land received	1.0	1.9	4.3	1.5	1.2	33.8	42.4	12.6	1.2	99.9
Dispersed per family (m^2)	----	----	----	1686	2049	2069	2005	1800	476	2060

Continued on next page

Table I - continued

	French	Church	Common	Land-lord	Rich Peas-ant	Middle Peas-ant	Poor Peas-ant	Land-less Peas-ant	Other Rural Work-ers	Total/ Average
1953										
Land used (owned +/- dispersed) (ha.)	144	277	630	4394	1954	20073	13021	2775	528	43796
Land used (owned +/- dispersed) per family (m^2)	----	----	----	33516	22881	8443	4241	2716	1413	6204
% of land used (owned +/- dispersed)	0.3	0.6	1.4	10.0	4.5	45.8	29.7	6.3	1.2	99.8
Land used (owned +/- landlord dispersed) (ha.)	420	687	9871	4175	1951	16314	8490	1507	381	43795
Land used (owned +/- landlord dispersed) per family (m^2)	----	----	----	31846	22845	6862	2765	1475	1020	6204
% of land used (owned +/-landlord dispersed)	1.0	1.6	22.5	9.5	4.5	37.3	19.4	3.4	0.9	100.1
% of common land used	----	----	6.4	2.2	1.5	35.4	41.6	11.3	1.4	99.8
% of landlord dispersed land used	----	----	----	----	0.6	34.0	47.7	16.7	0.9	99.9
Land used / owned	0.34	0.40	0.06	0.58	1.01	1.32	1.90	2.96	1.51	1.00
1945-53										
Change in land owned (ha.)	74	-5	-872	-2256	-1186	1888	2129	391	22	185
Change in land owned as %of 1945 owned	21.4	-0.7	-8.1	-22.9	-38.1	14.2	45.0	71.7	6.7	----
Land used 1953 / Land owned 1945	0.42	0.40	0.06	0.45	0.63	1.51	2.75	5.09	1.61	1.00[8]
Land used per family 1953 / Land owned per family 1945	----	----	----	0.68	0.94	1.17	1.81	5.88	1.50	0.81[8]

Notes to Table I

1 Average number of families per village was 18.3 in 1945 and 12.1 in 1953.

2 23.4% rise in the number of families, 1945-53 - 2.7% per annum.

3 0.4% rise in land owned, 1945-53.

4 18.6% fall in average land owned per family, 1945-53.

5 45% of landlord holdings were dispersed.

6 7.8% of rich peasant land was dispersed.

7 33.2% of the total land was dispersed.

8 The difference between these totals occurs because of a change in the percentage of families in the various categories and an increase in the total population.

Source - adapted from: Tran Phuong (1968) opp. p.82 and Hanoi (1960c) p.66 - data collected Post-Rectification from 93 villages and 31 hamlets.

Table II - Classification and Landownership, 1945-56

Classification	1945						During Land Reform				1953			1956
	Population (1)	% of Pop. (1)	% of Pop. (2)	% of Families (3)	% of land owned (1)	% of land owned (3)	% of Pop. (2)	% of Pop. (3)	% of Pop. (4)	% of Families (4)	% of land owned (3)	% of land used (2)*	% of land used (3)	% of land owned (2)
French	----	----	----	----	1.0[5]	0.8	----	----	----	----	0.9	----	0.3	----
Church	----	----	----	----	1.5	1.6	----	----	----	----	1.5	1.3	0.6	0.1
Communal	----	----	----	----	25.0[5]	24.6	----	----	----	----	22.5	4.3	1.4	0.7
Landlord (6) (7)	387,224	4.0	4.0	3.5	24.5	22.5	2.3	1.8	2.5	2.1	17.3	18.0	10.0	1.1
Rich Peasant (6)	284,931	2.9	2.9	2.2	7.1	7.1	1.6	1.2	1.7	1.2	4.4	4.7	4.5	1.8
Middle Peasant	3,371,789	34.8	34.8	32.2	29.0	30.4	36.5	33.6	36.2	31.9	34.6	39.0	45.8	41.7
Poor Peasant	3,701,172	38.2	37.0	35.4	10.0	10.8	42.9	43.5	42.7	44.1	15.6	25.4	29.7	40.5
Landless Peasant	1,415,101	14.6	17.1	20.6	1.1	1.2	13.1	14.4	13.0	16.7	2.1	6.3	6.3	12.0
Other Workers	540,720	5.6	4.2	6.1	0.8	0.7	3.6	5.3	3.5	3.7	0.8	1.0	1.2	2.6
Total	9,700,937	100.1	100	100	100	99.7	100	99.8	99.6	99.7	99.7	100	99.8	100.5

*owned land plus communal land

Notes and Sources to Table II

(1) Land Reform Committee cited Vo Nhan Tri (1967) and Tran Phuong (1968) p.14 - Post-Rectification data allegedly referring to 3,653 villages, though probably a smaller sample.

(2) Hanoi (1978) p.94 and Nguyen Xuan Lai (1976) - Pre-Rectification data which show a dramatic rise in poor peasant land 1953-56.

(3) Hanoi (1960c) p.66 and Tran Phuong (1968) p.32 and opp. p.82. Post-Rectification data from 93 villages and 31 hamlets in 16 provinces.

(4) Tran Phuong (1968) p.162 and Moise (1977) p.419. Post-Rectification data from 1,508 villages.

(5) Hanoi (1978) gives French 2.4%, Communal 23.6%.

(6) Tran Phuong (1968) p.156 and p.160 gives following landlord and rich peasant classifications.

	population		families	
	landlord	rich peasant	landlord	rich peasant
1945	3.9	2.9	3.4	2.2
1949	2.3	2.2	2.0	1.5
1953		1.6		1.1

(7) Moise (1977) p.419 and Tran Phuong (1968) p.158 give the breakdown of landlords as follows:

	population	% of total population	families	% of total families	% of families by category
despots	34,861	0.3	6,220	0.2	13.0
ordinary	185,399	1.7	37,411	1.4	78.1
resistance	23,476	0.2	4,269	0.2	8.9
	243,736	2.2	47,900	1.8	100.0

Table III - Collectivisation of Agriculture, 1955-75

Year	Indiv. Production			Low-level coops.			High-level Coops.	
	% of all families	% of cultivated land	Seasonal MATs	% of all families	% of cultivated land	Permanent MATs	% of all families	% of cultivated land
1955	59.5	....	37.2[1]	----	----	3.3[1]	----	----
1956	49.9	....	45.1	----	----	5.0	----	----
1957	78.1	....	19.1	----	----	2.0	----	----
1958	29.6	....	44.7	4.7	....	21.0	----	----
1959	17.1	....	26.2	43.0	35.0	11.3	2.4	2.0
1960	14.2	31.9		73.4	57.5		12.4	10.6
1961	11.1	29.1		58.3	46.7		30.6	24.2
1962	13.7	30.9		49.0	38.9		37.3	30.2
1963	14.9	31.9		45.9	37.1		39.2	31.0
1964	15.3	24.4		39.2	36.5		45.5	39.1
1965	9.9	19.7		25.1	25.9		65.0	54.4
1966	7.3	10.9		13.0	13.5		79.7	75.6
1967	5.9	9.1		12.7	12.7		81.4	78.2
1968	5.2	7.8		6.7	8.5		88.1	83.7
1969	4.9	7.6		5.4	7.8		89.7	84.6
1970	4.5	5.4		3.6	7.3		91.9	87.3
1971	4.1	5.0		3.9	7.5		92.0	87.5
1972	4.3	5.1		3.7	7.2		92.0	87.7
1973	5.2	5.3		3.5	7.1		91.3	87.6
1974	4.8	5.0		2.6	6.6		92.6	88.4
1975	4.4	4.8		2.5	6.5		93.1	88.7

Note: 1 percentages of families

Source: Derived from Hanoi (1960e) pp. 84-86, (1977c) p. 109, and (1978) pp.97-98.

Table IV - Agricultural Production, 1954-75

	1954	1955	1956	1957	1958	1959	1960	1961	1962	1963	1964	1965	1966	1967	1968	1969	1970	1971	1972	1973	1974	1975
Population (mid-year)	13.4	13.8	14.2	14.6	15.0	15.5	16.0	16.6	17.2	17.7	18.3	18.9	19.5	20.1	20.7	21.2	21.8	22.4	22.9	23.5	24.0	24.6
Paddy Production (m tons)	2.55	3.52	4.13	3.95	4.58	5.19	4.18	4.39	4.39	4.11	4.43	4.55	4.13	4.29	3.71	3.91	4.46	4.12	4.92	4.47	5.49	4.78
Paddy per capita (kg)	190	256	292	271	304	335	261	281	256	232	243	240	212	214	179	184	204	184	215	190	228	194
Maize Production (m tons)	0.15	0.19	0.24	0.20	0.19	0.25	0.22	0.26	0.26	0.22	0.28	0.28	0.22	0.26	0.20	0.19	0.23	0.20	0.22	0.20	0.23	0.21
Potato Production (m tons)	0.27	0.53	0.87	0.54	0.42	0.72	0.57	0.95	0.82	0.85	1.18	1.19	1.24	1.48	1.12	1.02	0.94	0.90	0.86	0.75	0.73	0.65
Manioc Production (m tons)	0.13	0.16	0.37	0.19	0.15	0.26	0.33	0.70	0.75	0.89	0.92	0.80	0.81	0.88	0.71	0.60	0.64	0.70	0.75	0.63	0.75	0.67
Subsidiary Production (paddy equiv. m tons)	0.28	0.42	0.65	0.44	0.38	0.57	0.52	0.81	0.79	0.90	1.09	1.02	0.97	1.11	0.92	0.80	0.82	0.80	0.82	0.72	0.79	0.71
Subsidiary per capita (paddy kg)	21	30	46	30	26	37	33	49	46	51	60	54	50	55	44	38	38	36	36	31	33	29
Total Crop Production (m tons)	2.83	3.94	4.79	4.39	4.96	5.77	4.70	5.20	5.17	5.01	5.52	5.56	5.10	5.40	4.63	4.71	5.28	4.92	5.74	5.19	6.28	5.50
Total Crop per capita (kg)	211	286	337	301	331	372	293	314	301	283	302	294	262	269	224	222	242	220	250	221	261	224
Vegetable & Beans Production (m tons)	0.15	0.16	0.17	0.18	0.28	0.31	0.32	0.42	0.49	0.63	0.70	0.82	0.84	1.05	1.15	1.13	1.52	1.35	1.51	1.39	1.54	1.47
Vegetable & Beans per capita (kg)	11	12	12	12	19	20	20	25	28	36	38	43	43	52	56	53	70	60	66	59	64	60
Paddy acreage (m ha)	1.86	2.18	2.28	2.19	2.24	2.27	2.27	2.39	2.41	2.36	2.43	2.40	2.39	2.19	2.08	2.15	2.21	2.07	2.20	2.09	2.27	2.25
Paddy yield (ta/ha)	13.7	16.1	18.1	18.0	20.4	22.9	18.4	18.4	18.2	17.4	18.2	19.0	17.3	19.6	17.8	18.2	20.2	19.9	22.4	21.4	24.2	21.2
Subsidiary acreage (m ha)	0.23	0.34	0.35	0.30	0.32	0.33	0.36	0.49	0.52	0.59	0.64	0.61	0.63	0.66	0.58	0.52	0.51	0.51	0.50	0.47	0.48	0.46
Subsidiary yield (paddy equivalent ta/ha)	12.2	12.4	18.6	14.7	11.9	17.3	14.6	16.5	15.2	15.3	17.0	16.5	15.5	16.8	15.9	15.3	16.1	15.5	16.3	15.4	16.5	15.4
Vegetable acreage (m ha)	.014	.015	.015	.016	.026	.029	.031	.040	.048	.062	.068	.074	.086	.093	.097	.107	.113	.105	.117	.116	.115	.116
Vegetable yield (ta/ha)	← 100 used as basis for output etc →											108	96	111	117	105	134	127	128	119	132	127
Structure of annual cultivated area (%s)																						
Paddy	85.2	82.1	82.1	82.2	81.7	81.2	79.8	76.9	76.6	72.7	72.2	71.9	71.7	69.8	70.7	72.5	73.1	71.9	73.1	73.4	74.7	74.6
Subsidiary Crops	10.5	12.8	12.6	11.3	11.7	11.8	12.6	15.8	16.2	18.1	19.0	18.4	18.8	20.8	19.7	17.6	16.9	17.9	16.8	16.5	15.8	15.4
Vegetables & Beans	2.0	2.5	2.3	2.6	2.5	2.8	3.3	3.2	3.3	3.8	3.7	3.9	4.1	4.5	4.8	5.1	5.1	5.0	5.2	5.3	5.0	5.0
Industrial Crops	2.3	2.6	3.0	3.9	4.1	4.2	4.2	4.0	4.7	5.0	4.8	5.4	5.1	5.0	4.7	4.6	4.8	4.9	4.6	4.5	4.2	4.5

Notes to Table IV

1. The population series is derived from 1960 and 1974 census data using estimated growth rates. See also Ng Shui Meng (1974).

2. All data for 1960-75 except population are taken from Hanoi (1977c) or calculated on the basis of data therein.

3. Pre-1960 data are based on official data cited in the following sources: Hanoi (1960c, d,f, and 1978); Vo Nhan Tri (1967); Le Chau (1966); Lavallèe (1971); Nguyen Tien Hung (1977); Karamyshev, Vietnamese Studies (1959); and in BBC SWB. They have been cross-checked for consistency. Subsidiary crop production data have been standardised using conversion ratios in Hanoi (1977c) as follows:
1kg paddy = 1kg maize = 3kg manioc = 3kg potatoes.
The subsidiary crop production total increasingly diverges from the sum of these three components because of the increasing output of other crops. The pre-1960 figures, which do not take this into account, may hence be underestimates.

4. Vegetable and bean production figures have been added together and exclude soya beans. The pre-1965 output figures for this are based on area data assuming productivity of 100 quintals per hectare. This and the area figures may be an underestimate; if so, the output increase is exaggerated.

5. Discrepancies between this and other series are likely to result from a) the population series used, b) the conversion ratios for subsidiary crops, c) the inclusion of beans in subsidiary crop figures, and d) inadequate cross-checking of output, area, and yield data in other series.

6. A number of other reservations in using or interpreting the data are noted in the text.

Table V - Livestock 1955-75

	1955	1956	1957	1958	1959	1960	1961	1962	1963	1964	1965
Water buffalo (million)	1.04	1.12	1.25	1.39	1.45	1.45	1.43	1.46	1.51	1.57	1.61
Oxen (million)	0.68	0.78	0.90	1.05	0.95	0.86	0.76	0.78	0.70	0.82	0.81
Pigs (million)	2.20	2.50	2.95	3.91	3.63	3.81	3.81	4.42	4.46	4.56	4.79
Horses (thousand)	20.0	21.0	24.0	24.0	....	39.4	42.2	45.8	49.8	56.1	57.7
Goats (thousand)	....	....	....	....	42.9	48.8	57.3	60.7	68.7	99.0	114.5
Poultry	10.8	14.2	15.9	17.7	....	20.4	21.2	21.6	32.7	23.3	30.0

	1966	1967	1968	1969	1970	1971	1972	1973	1974	1975
Water buffalo (million)	1.66	1.64	1.62	1.65	1.71	1.73	1.73	1.77	1.76	1.80
Oxen (million)	0.79	0.76	0.72	0.72	0.71	0.69	0.68	0.65	0.64	0.66
Pigs (million)	5.10	5.10	5.27	5.30	5.38	5.63	5.73	6.33	6.41	6.60
Horses (thousand)	60.4	66.4	84.8	93.8	102.3	107.5	113.3	118.7	124.1	124.8
Goats (thousand)	121.0	120.4	113.2	132.3	134.2	130.1	143.8	135.6	138.7	147.4
Poultry	27.2	28.0	28.8	28.9	30.4	28.9	33.7	34.9	33.8	35.7

Note: See also collective pigs date in Nguyen Xuan Lai and Vu Quoc Tuan (1976) p.184.

Sources: Karamyshev (1959) p.47, Nguyen Chi Thanh (1963a) p.230, and Hanoi (1977c) p.131.

Note to Table V

Vo Nhan Tri (1967) pp.218,295,429 gives the following figures in '000s:

	Buffalo	Oxen	Pigs
1939	788	563	2255
1955	1084	756	2137
1956	1165	834	2500
1957	1238	906	2950
1958	1376	992	3985
1959	1447	950	3626
1960	1451	879	3751
1961	1438	781	3740
1962	1480	786	4239
1963	1508	796	4209
1964	1565	819	4528

Table VI - State Share of Food Production, 1955-75

Year	Tax & Purchases				State[1] Wholesale				State[1] Retail				Traded share of value of agri. prod.	Maize procurements (000 tons)	Meat procurements (000 tons)	Retailed Meat	Wholesale Meat
	Paddy (000 tons)	%[2]	Staples	%[2]	Paddy	%[2]	Staples	%[2]	Paddy	%[2]	Staples	%[2]					
1955	682	19.4	...	...	...	...	...	...	310[3]	13.5	...	...	...	6.9	...	...	...
1956	534	13.0	...	...	...	...	...	...	...	...	...	...	...	...	...	...	...
1957	590	15.0	...	...	...	...	...	...	344[3]	13.4	...	...	...	39.5	6.4	...	...
1958	674	14.7	...	...	...	...	...	...	...	...	...	...	...	35.5	30.5	...	...
1959	853	16.4	...	...	...	...	...	...	...	...	...	...	...	52.1	70.8	...	...
1960	782	18.7	878	18.7	834	20.0	868	18.5	581[3]	21.4	468	10.0	15.4	32.3	96.0	26.5	67.5
1961	804	18.3	...	...	...	...	...	...	...	...	...	...	14.3	28.2	56.5	...	...
1962	890	20.3	...	...	...	...	...	...	...	...	...	...	16.8	29.9	59.9	...	...
1963	695	16.9	...	...	...	...	...	...	...	...	...	...	14.8	18.0	56.0	...	...
1964	930	21.0[5]	...	...	...	...	...	...	593[3]	20.6	601	10.9	18.3	30.9	73.3	19.5	...
1965[4]	1058	23.3	1125	20.2	1048	23.0	1106	19.9	...	...	684	12.3	19.9	...	78.0	43.4	90.0
1966	...	...	...	...	...	...	...	...	...	...	797	15.6	15.2	...	...	...	...
1967	...	...	...	...	...	...	...	...	...	...	871	16.1	17.5	...	...	...	...
1968	...	...	...	...	...	...	...	...	...	...	930	20.1	16.4	...	...	45.1	...
1969	...	...	...	...	...	...	...	...	...	...	...	...	15.5	...	...	...	...
1970	...	...	...	...	...	...	...	...	...	...	...	...	13.8	...	...	...	...
1971	...	...	...	...	...	...	...	...	...	...	...	...	20.7	...	...	...	...
1972	...	...	...	...	...	...	...	...	...	...	1165	20.3	22.6	...	...	60.1	...
1973	...	...	...	...	...	...	...	...	...	...	...	...	21.2	...	...	...	...
1974	982	17.9	1017	16.2	979	17.8	1009	16.1	...	...	1278	20.4	19.8	...	103.7	58.8	98.2
1975	762	15.9	796	14.5	766	16.0	787	14.3	...	...	1193	21.7	24.4	...	122.9	70.0	118.3

Notes to Table VI

1. State includes cooperative stores.

2. Percentages are of the total harvest.

3. Figures for rice.

4. Burchett (1965) states 23% of harvest went into reserves.

5. Charrière (1964) states tax share alone as 8% and above quota sales as 3%.

Sources: Le Chau (1966); Vo Nhan Tri (1967); Hanoi 1977c); and Nguyen Xuan Lai and Vu Quoc Tuan (1976).

Additional Notes

1. White (1982) p.19 claims taxes and purchases accounted for 25-30% of cooperative paddy production and 10% of private production.

2. Charrière (1966) pp. 31-32 states obligatory deliveries as 12-24% of staples production.

3. Elliot (1976) p.518 gives procurement ratios for paddy from two cooperatives as follows:

1960	- 21.5	....
1961	- 19.5	20.6
1962	- 25.1	16.0
1963	-	17.0

4. Nguyen Xuan Lai (1967) gives example where tax represents 8.7% and quota sales 13.4% of paddy allocation to a cooperative family.

5. Nguyen Xuan Lai and Vu Quoc Tuan (1976) state compulsory deliveries average 15% of production during 1961-1964 and 20% during the subsequent war period.

Table VII - Growth and Structure of Gross Value of Agricultural Output, 1955-75

(at constant prices)

Year	Index	Annual Change	Cultivation	Livestock
1955	75.5	....	82.5	17.5
1956	88.9	17.7	....	....
1957	87.6	-1.5	83.9	16.1
1958	102.3	16.8	....	....
1959	110.7	8.2	....	....
1960	100.0	-9.7	79.1	20.9
1961	104.8	4.8	81.1	18.9
1962	111.7	6.6	79.3	20.7
1963	109.2	-2.2	78.0	22.0
1964	117.7	7.8	78.2	21.8
1965	123.1	4.6	77.1	22.9
1966	118.2	-4.0	77.6	22.4
1967	121.0	2.4	78.9	21.1
1968	110.0	-9.1	77.2	22.8
1969	113.1	2.8	76.3	23.7
1970	122.5	8.3	80.7	19.3
1971	118.3	-3.4	78.8	21.2
1972	132.9	12.3	80.1	19.9
1973	126.2	-5.0	77.3	22.7
1974	137.2	8.7	80.0	20.0
1975	130.5	-4.9	75.8	24.2

Note: Agricultural production accounted for 85-90% of the total value of agriculture, forestry, and other peasant production.

Sources: Truong Van Ba (1962) p.118
Nguyen Xuan Lai (1971)
Hanoi (1977c) p.111
Lavallee (1971) p.51
See also Van Dyke (1972) pp.190-91.

Table VIII - Value of Labour Productivity

Year	Workers in Agriculture (000)	Gross Value of Agricultural Production (m. dong)	Labour Productivity	
			Dong per worker	Index (1960=100)
1960	5759	2327.1	404	100
1965	5987	2863.7	478	118
1968	4928	2559.4	519	128
1969	4980	2633.0	529	131
1970	4910	2850.4	581	144
1971	5165	2753.0	533	132
1972	5269	3092.9	587	145
1973	5367	2937.6	547	135
1974	5654	3192.8	565	140
1975	5758	3036.4	527	130

Note: The fall in the number of workers in agriculture occurs despite the rural workforce remaining an approximately constant percentage of the total workforce.

Source: Hanoi (1977c) pp.111 and 112.

Year	% of Labour Force in Agriculture	% of Rural Population of Working Age	Agricultural workforce as % of Rural Population of Working Age
1960	83.0	87.5	87.8
1965	78.6	86.8	82.1
1975	64.7	84.7	62.1

Source: Hanoi (1977c) Tables 10 and 11, pp. 43-44.

Table IX - Productivity in Cooperatives

A. Workdays devoted to collective production

Year	Number	Value of Workday
1959	90[1]	
1960	103[1,3]	0.8[3]
1961	124[1,3] 135[6]	0.75[1] 0.77[3]
1962	140[1] 100[2] 130[4]	0.75[1]
1963	163[1]	
1964	160[1] 150[4,5] 100-135[7]	1 dong[1]

Sources:
1. Nhan Dan 25/9/64 cited by Elliot (1976) p.490
2. Burchett (1966)
3. Nguyen Chi Thanh (1963a) p.232
4. Gordon (1981) p.36
5. Charrière (1966) p.40
6. Le Duan VNA 16/8/62, SWB WS No. 176, p.B18
7. Dumesnil (1970) p.529

B. Targets (Work Days)

i. 220 to cooperative + 25 to State

ii. 200 to cooperative + 50 to labour service + 50 to family

iii. 250 to cooperative

i. Nguyen Xuan Lai (1967)
ii. Nhan Dan cited by Elliot (1976) pp.497-98
iii. H 7/10/65, SWB WS No.337, p.A40

Continued on next page

Table IX - continued

C. Data from three 'model' cooperatives

(1)	(2)	(3)	(4)	(5)	(6)	(7)	(8)	(9)
52.24	1045	403	7.7	13.0	427	8.2	618	1.53
71.00	1420	511	7.2	13.9	598	8.4	822	1.61
40.27	805	519	12.9	7.8	438	10.9	367	0.71

(1) Output - quintals per hectare

(2) Output - gross value in dong valuing output at 0.2 dong per kg

(3) Workdays per hectare

(4) Workdays per quintal

(5) Productivity kg per day

(6) Non-labour costs - dong per hectare

(7) Non-labour costs - dong per quintal

(8) Value less non-labour costs

(9) Value of workday in dong

D. Productivity

i. 120-140 working hours required to produce 100kg of paddy equivalent to approximately 5kg per day.

ii. 2500kg of paddy requires 500 workdays equivalent to 5kg per day.

iii. 2500kg of paddy requires 333 workdays equivalent to 7.5kg per day.

iv. 100kg of paddy requires 15 to 16 workdays equivalent to 6.5kg per day. Each peasant cultivated only 0.53 hectares.

i. Nguyen Xuan Lai (1971)
ii. Dumont (1965a)
iii. Nhan Dan 28/7/59 cited by Elliot (1976) p.456.
iv. Nhan Dan 18/12/71, SWB WS No. 612, p.All.

Table X - Peasant Income Data

(dong per capita per annum)

1961	average	138[1]	120[2]
	poor	129[1]	
	middle	149[1]	
	upper-middle		160[2]
	rich	181[1]	
1962	average	145[2]	
1958-60	average	146[3]	
1963	average	192-216[4]	

Sources:
1 Nguyen Chi Thanh (1963a) p.232
2 Elliot (1976) p.498
3 Le Chau (1967) p.357
4 Charrière (1966) p.33

1960 A. 103 days x 0.80 dong per day[1] = 82.4 dong
Family had 1.84 workers[2] = 152 dong
Cooperative contributed 45% of income[1]
∴ Total income = 337 dong
average family size = 4.1 members[3]

∴ Income per head = 82 dong

B. Cooperative contributed 35% of income[4]
∴ Total income = 434 dong
= 106 dong per capita

Sources:
1 Nguyen Chi Thanh (1963a) p.232
2 Hanoi (1977c) p.111
3 Land Reform data, see Table II
4 Hanoi (1977c) p.180

1967 Nguyen Xuan Lai (1967) gives data on a cooperative family with seven members as follows:

Receives 2200kg of paddy = 572 dong

Other income = 455 dong (excludes consumption of own vegetables)

Total = 1027 dong i.e. 147 dong per capita

Table XI - Industrial and Handicraft Production Indexes, 1955-75

Year	Gross output[1]	Fixed capital[2]	Work force[2]	Dept. A gross output	Dept. B gross output	Dept. A share	Dept. B share	Industrial gross output[1]	Handicraft gross output[1]	Industrial share	Handicraft share	Central gross output[1]	Local gross output[1]	Central share	Local share	Gross output per capita[3]	Dept. B output per capita[4]	Cotton cloth output[5]
1955	22	...	...	17	24	27	73	11	36	29	71	7	31	12	88	23	28	12
1956	43	...	...	36	46	29	71	31	58	41	59	27	53	25	75	44	52	63
1957	57	...	...	38	67	24	76	41	77	43	57	42	66	28	72	59	73	83
1958	69	...	...	51	78	25	75	58	83	48	52	64	72	36	64	68	83	87
1959	89	...	...	85	92	32	68	84	95	54	46	107	77	48	52	85	95	103
1960	100	100	100	100	100	34	66	100	100	57	43	100	100	40	60	100	100	100
1961	105[2]	114	105	...	...	...	...	...	101[2]	...	...	...	...	...	...	110	...	...
1962	143	133	110	157	136	37	63	163	116	65	35	168	123	47	53	133	127	125
1963	154	157	113	178	141	39	61	181	118	67	33	187	130	48	51	141	127	105
1964	178	182	122	208	162	40	60	218	124	70	30	233	141	52	48	152	142	141
1965	194	216	127	242	169	43	57	250	119	74	26	264	148	54	46	160	143	132
1966	188	252	124	254	154	46	54	237	124	72	28	241	152	51	49	151	126	...
1967	164	276	122	209	140	44	56	199	116	69	31	187	148	45	55	127	111	...
1968	173	298	122	228	145	45	55	221	110	73	27	205	152	47	53	131	112	109
1969	185[2]	324	126	...	...	...	...	...	116[2]	...	...	...	...	...	...	139	...	...
1970	200[2]	348	139	...	...	...	...	...	122[2]	...	...	...	...	...	...	147	...	...
1971	237	373	154	287	210	41	59	316	131	76	24	294	199	49	51	166	150	143
1972	210	396	151	248	190	40	60	271	129	74	26	241	190	45	55	144	133	96
1973	250	417	156	306	221	42	58	321	155	73	27	279	231	44	56	165	150	100
1974	291[2]	438	172	376[2]	249[2]	43[2]	57[2]	...	179[2]	...	...	343[2]	257[2]	45	55	194	166	...
1975	336	448	178	439[2]	284[2]	43[2]	57[2]	...	201[2]	...	...	427	276	51	49	219	185	116

Continued on next page

Notes and sources to Table XI

All data from official sources cited in Nguyen Tien Hung (1977) p.140 except as below:

1 At constant (1959) prices.

2 From Hanoi (1977c) Tables 30-32, pp.65-67 and Tables 43-45, pp.77-79. Also at constant 1970 prices, these are sufficiently compatible to the other data.

3 Derived from Population Series (Table IX) and gross output series constructed from data in Van Dyke (1972) p.190, Nguyen Xuan Lai (1971), Le Vinh (1962), pp.105-6, Hanoi (1977c) Table 30, p.65.

4 Using Dept.B Gross Output series and Population Series (Table IX).

5 Official sources cited in Nguyen Tien Hung (1977), p.144.

Notes to Introduction

1. See, for example, the summary of Taiwan's experience in Hsieh (1971).

2. The only existing attempts to do this are the works of James C. Scott (1976) and Samuel Popkin (1979) who employ 'moral economy' and 'rational peasant' arguments respectively.

3. Eckstein (1968), p.508.

4. Tang (1968), p.460.

5. Millar (1970b), p.78.

6. Baran (1973), pp.424-25. See also Charriere (1965), p.39.

7. For this debate, see Dobb (1966) and Erlich (1967); for some of the original contributions see Preobrazhensky (1965) and Bukharin (1982).

8. From Preobrazhensky's 'self-criticism' at the 17th Congress of the CPSU in 1934 quoted in Nove (1965), p.xv.

9. The notion of one sector 'financing' another and the definitions of surplus are more complex than they appear. For a discussion of this see Millar (1970b), pp.78-82.

10. Nove (1965), p.xi. Chayanov's analysis suggests that adverse terms of trade would be effective in raising marketings. See Millar (1970a), p.229.

11. For a detailed exposition of Preobrazhensky's views see Preobrazhensky (1965) and Ehrlich (1950).

12. For Bukharin's argument and his criticism of Preobrazhensky, see Bukharin (1982).

13. Bukharin (1982), p.310.

14. See the following discussion on whether

or not agriculture constrains industry; also Limqueco and McFarlane (1980), p.52.

15. Cited in Nove (1965), p.xii.

16. There is continual debate over the nature of peasant responses to changes in the terms of trade. See, for example, Millar (1970a), p.29.

17. Bergson cited in Tang (1968), p.460.

18. Cited in Nove (1965), p.xiv.

19. Millar (1970b), (1971) and Ellman (1975) using Soviet data. For some disagreement with their argument see Nove (1971).

20. Ellman (1978).

21. "The labour transfer from agriculture to industry was the major contributing factor to economic development on the part of the agricultural sector." Kahan (1964), p.264.

22. Ellman (1975), p.859.

23. Limqueco and MacFarlane (1980), p.53.

24. See Ellman (1975), p.860, footnote 2.

25. Walker (1966), for example, pp.14-16.

26. Paine (1976), p.285.

27. Collectivisation in South Vietnam post-1975 was partly rationalised on the grounds that "whereas taxation on individual landholdings was only some 10% of the harvest, with collectivisation the percentage procured could be raised to 30-40%," Vo Chi Chung cited in White (1982a), p.19.

28. If 'primitive accumulation' or 'original socialist accumulation' (Bukharin (1982)) is seen as the transfer of workers then it did occur, to a degree, in all three countries. The focus here is on the financing of accumulation which is normally seen as the key attribute of the Soviet model. See the discussions in Millar (1970b), Baran (1973),

especially pp.416-42, and also Dobb (1966), Ch.8.

29. Mao clearly believed that agriculture <u>had</u> financed a rapid rate of industrialisation in the USSR. Walker (1966), p.31. It had also been the <u>intention</u> in the USSR Nove (1971), pp.394-95.

30. Millar (1970b), p.93.

31. Tang (1968).

32. Ellman (1975), p.859.

33. Tang (1968), p.470.

34. Ibid. For China see also data in Paine (1976).

35. Available data on procurement is presented in Appendix Table VI. The range of sources used for this table, its gaps and inconsistencies, make it impossible to draw any firm conclusions from it.

36. Derived from Ellman (1975), p.847, Table 3.

37. Nolan (1976), p.199, Table 6.

38. Appendix Table VI.

39. Paine (1976), p.285.

40. See Walker (1966) for an excellent summary of the arguments for collectivisation in China. See Kahan (1964) for the arguments in the USSR, and Chapter IV below, for the DRV.

41. Kautsky cited by Wittfogel (1971), p.17. For this classical presentation and some inherent contradictions see Kautsky (1900).

42. This amendment of orthodoxy can really be traced to Lenin. See Wittfogel (1971) for a discussion of the evolution of this argument.

43. Walker (1966), p.7. He discusses the Chinese view in detail, both pre-1954 when it broadly accepted the Stalin argument and post-

1954.

44. Ibid. p.3.

45. The argument is flawed at a theoretical level since it fails to account for certain crucial characteristics of agricultural production and related peasant attitudes. These are further discussed below.

46. Walker (1966), p.7 (emphasis in original).

47. Perkins (1969), p.189.

48. Ishikawa (1967), p.122.

49. Ibid.

50. In particular, Perkins (1969), p.8 who suggests that the Chinese leadership also recognised this. See also Tang (1968). Evidence from Taiwan, Hsieh (1971), would also tend to confirm Ishikawa's hypotheses.

51. Tang (1968), p.496.

52. See below, also Ishikawa (1967), p.257.

53. Nolan (1976), p.208.

54. The same argument has of course been made for China. See, for example, Walker (1965), (1968), and Wittfogel (1971).

55. See, for example, Ishikawa (1967) and Hsieh (1971).

56. This point has also been made in more general terms by Bradley and Clark (1972) who suggest that "changes in agricultural technology would tend to be labour-saving in the absence of changes in relative factor prices because of the high cost of supervising agricultural labour." p.473.

57. Marx and Engels cited by Baran (1973), p.434. Baran's own comments following this quotation refer correctly and precisely to the

problems of collectives. However, his belief in their merit as a means of extracting a surplus overrides these concerns.

58. Bradley and Clark (1972), p.469.

59. Ibid. pp.466-8. Similar points are made by Wittfogel (1971) especially pp.22-29.

60. For supervision and other related difficulties in China, see Walker (1968).

61. Nguyen Khac Vien in Gough (1982), p.378.

62. Bradley and Clark (1972), pp.470-1.

63. See, for example, Walker (1965) and (1968), pp.426-31.

64. See the citations from Engels in Mandel (1968), p.647.

65. Chayanov (1966). For a concise summary of Chayanov's contribution and some weaknesses see Kerblay (1971).

66. Thorner (1966), p.xviii.

67. Harrison (1975), p.390.

68. See Kerblay (1966), pp.lx-lxiv.

69. Ibid. p.lix, see also p.xliv.

70. Millar (1970a), p.220.

71. Ibid. See also Harrison (1975), p.398.

72. Chayanov (1966), pp.111-12.

73. Millar (1970a), p.221.

74. Thorner (1966), p.xxii. See also, Kerblay (1966), pp.lxix-lxx.

75. Harrison (1975), p.416.

76. See, for example, Shanin (1972, 1973) and Littlejohn (1973a,b). The best critical apprais-

als of Chayanov are Harrison (1975) and Millar (1970a). See also, Kitching (1982), Ch.3.

77. For a critique of 'snapshots' of peasant differentiation such as was presented by the DRV leadership see Shanin (1980), p.96.

78. Chayanov (1966), p.41. Harrison (1975) considers Chayanov's analysis of labour allocation in detail.

79. Cf. the discussion of the terms of trade above and Ellman (1975), pp.849-50.

80. Cf. the discussion in Ishikawa (1967), pp.249-67.

81. Aspects of this relationship between the sectors is considered in Chapter V. See Kerblay (1980) for an excellent discussion of the Kolkhoz/ family plot relationship in the USSR.

82. However, see the different arguments on the effect of terms of trade changes on peasant marketings where incomes are low, Millar (1970a), pp.225-29, also Millar (1971), p.86.

83. For a belated recognition of both of these points, see Nguyen Khac Vien in Gough (1982).

84. Nolan (1976), p.215.

Notes to Chapter I

1. The most detailed western work on the geography and ecology of the Red River delta remains Gourou (1936). See also Fridland (1961) and Henry (1932).

2. Appendix Table IV gives the breakdown of the cultivated area for 1954-75.

3. For information on local administration in the DRV both during the resistance war and after 1954, see Ginsburgs (1962 and 1963).

4. Mau (1977).

5. Chesneaux (1955), pp.72-73.

6. Woodside (1971) and Cheshkov (1969).

7. For a stylised description of the pre-colonial commune and land distribution within it, see Adams and Hancock (1970); also Ngo Vinh Long (1973), pp.6-9.

8. Le Thanh Khoi (1955), Chesneaux (1955), and Hodgkin (1981) are the major works available in the West on Vietnamese history.

9. Woodside (1971), p.135 and Ngo Vinh Long (1973), pp.31-32.

10. Cheskhov (1969) and Ngo Vinh Long (1973).

11. For a discussion of 'class' development in Vietnam, see Dumarest (1935).

12. Perkins (1969) makes a similar argument for China in the pre-modern period.

13. On the last pre-colonial regime, the Nguyen dynasty, see Buttinger (1958) Ch.V, Chesneaux (1955), and Woodside (1971).

14. Marr (1981), p.3.

15. Thompson (1937), p.212. The main western sources on the colonial economy in general are Robequain (1944), Thompson (1937), Bernard (1934), and Murray (1980).

16. For information on the 'working-class' see Le Thanh Khoi (1955), Goudal (1938), Murray (1981), and Hémery (1975).

17. Buttinger (1967).

18. Chesneaux (1962), p.23 and Nguyen Khac Vien (1964), p.3.

19. Nguyen Tien Hung (1977), p.9.

20. Food and Agricultural Organization (FAO) (1948).

21. Annuaire Statistique (1942).

22. FAO (1948).

23. Ishikawa (1967), p.70.

24. Particularly Henry (1932) and Gourou (1936).

25. See Appendix Table IV.

26. Perkins (1969), p.11.

27. Gourou (1936).

28. Walker (1968), pp.405-17.

29. United States Department of Agriculture (1950), p.16.

30. Ibid., p.17.

31. See the later discussions of irrigation. Tang (1968) analyzes the limited contribution of further labour intensive irrigation projects in China in the 1950s.

32. Perkins (1969), p.189.

33. Ishikawa (1967), particularly Chapter 2, considers the relative contribution of different inputs. As argued later in this volume and by others such as Perkins (1969), water is the significant input up to a point after which fertiliser, increasingly of the chemical kind, and other 'modern' inputs become the major contributors to increased production.

34. For an assertion that this _is_ the case, see Nolan (1976), p.202. Tang (1968), pp.491-503 provides evidence on the limited extent to which labour mobilisation in China produced returns.

35. For a discussion of the 'march southwards' see Le Thanh Khoi (1955), pp.163-69, Buttinger (1958), Ch.I, and (1967) note 3, Ch.IV.

36. This table is adapted from Henry (1932) which is effectively the sole Western source of data on the colonial land tenure. The figures indicate orders of magnitude since there were considerable variations between provinces. The data is used in almost all literature on this aspect of the colonial economy. See for example, Tran Phuong (1968), pp.9-13. Gourou (1936), the other major Western work on colonial agriculture, makes a number of searching criticisms on Henry (1932).

37. Nguyen Xuan Lai (1976), p.18.

38. Tran Phuong (1968), p.212.

39. See Truong Chinh and Vo Nguyen Giap (1974), pp.82-83, and Dumarest (1935) for a description of the alienation of communal lands by the landlords and rich peasants.

40. Chayanov (1966). See also Harrison (1975), p.403, and Miller (1970a), p.227.

41. Thompson (1947), p.6. For detailed information on moneylending and the extremely high interest rates, see Truong Chinh and Vo Nguyen Giap (1974), Ngo Vinh Long (1973), Henry (1932), and Gourou (1936 and 1945).

42. Murray (1980) sees the existence of 'small-scale petty proprietorship' as _itself_ the retarding factor, rather than the location of this system (if it is even an adequate description) _within_ the colonial economy.

43. Truong Chinh and Vo Nguyen Giap (1974), p.43.

44. Cited by Buttinger (1967), p.57.

45. See McAllister (1969), p.76, Mus (1949), p.269, Bernard, P. (1934), Buttinger (1967), p.56, Roberts (1929), p.481, Scott (1976), p.106. For tax details and examples see Truong Chinh and Vo Nguyen Giap (1974), Ngo Vinh Long (1973), and

Gourou (1936).

46. Thompson (1937) and Chesneaux (1955).

47. Marr (1981), p.29.

48. Bernard (1934), p.25

49. See, for example, McAllister and Mus (1970), McAllister (1969), Marr (1971 and 1981), Duiker (1975b) and, in the context of the war with the US, Fitzgerald (1973).

50. Delattre (1961), p.16.

51. Le Vinh (1962), p.104.

52. Western and Soviet sources cited by Nguyen Tien Hung (1977), p.28. This provides a detailed discussion of trading links and 'economic complementarity' between the two parts prior to partition, see pp.3-37.

53. Vo Nhan Tri (1967), p.261.

54. Ibid., p.260.

55. Charrière (1966), p.22.

56. Le Chau (1966), p.237.

57. For the Three-Year Plan see Hanoi (1959), and Nguyen Duy Trinh (1958).

58. Nguyen Tien Hung (1977), p.82.

59. Hanoi (1959), p.120.

60. Nguyen Tien Hung (1977), p.90.

61. Vo Nhan Tri (1967), p.257, Le Chau (1966), pp.245-47.

62. Le Chau (1966), p.205 and Vo Nhan Tri (1967), p.257.

63. Nguyen Xuan Lai (1971), pp.150-51; see also Charrière (1966), p.40.

64. Hanoi (1977c) Table 26, p.61.

65. Ibid., Tables 11 and 73, pp.43 and 112.

66. Charrière (1966), p.39.

67. This was discussed in the Introduction. See also Tang (1968), who discusses the significance of the constraint in the cases of China and the USSR.

68. Nguyen Tien Hung (1977), p.142.

69. The data in this and the following paragraph is presented in Appendix Table XI and in Figures 3-7. For a discussion of local industry in China see Riskin (1971).

70. Nguyen Xuan Lai (1971), p.142 and p.165. 'L'industrie légère a pour tâche de faire valoriser les produits agricole,' Vo Nhan Tri (1967), p.384.

71. Nguyen Khac Vien in Gough (1982), p.378.

72. Nhan Dan 11/8/68, SWB WS No. 482, p.A15. See also Vo Nhan Tri (1967), p.383, Nguyen Xuan Lai (1971), p.142, and Van Dyke (1972), pp.167-69 who cites some contemporary comments. For similar comments on agricultural equipment in China see Riskin (1971), p.255.

73. See, for example, Charrière (1966), pp.34-37.

74. Hanoi (1977c) Table 21, p.56.

75. For the early development of the ICP from varying perspectives see Duiker (1975b,c and 1981), Turner (1975), Pike (1978), Hanoi (1976), Rousset (1975), Porter (1976), and Hodgkin (1981).

76. For a discussion of the varying influences on the ICP/VWP see Turley, ed. (1980).

77. For an account of the Nghe-Tinh Soviets see Tran Huy Lieu (1980), also Scott (1976), and Sampson (1975).

78. Hodgkin (1981), pp.267-75. The 1935 Macao Congress which had confirmed a narrow 'class line' was not attended by Ho Chi Minh. For the relationship between the ICP and the Comintern see Duiker (1975a).

79. The best English language biography of Ho Chi Minh is Lacoutre (1978). In addition, the references cited in footnote 75 contain much useful information.

80. Ho Chi Minh (1960) vol.I, p.156.

81. For Ho Chi Minh's role in the pre-1930 Communist movements in Vietnam see Porter (1976).

82. Duiker (1975b), pp.266-70.

83. On this Plenum see Hanoi (1960d), and Duiker (1975b), pp.250-57.

84. See White (1979), pp.28-30.

85. For analyses of the 1945-46 period see Buttinger (1967), and Devillers (1952), and for many of the important documents Cole, ed. (1956), and Porter, ed. (1979).

86. For the development of agrarian policy before and during the resistance period see Tran Phuong (1968), Moise (1977), and Chapter II of this volume.

Notes to Chapter II

1. Nguyen Khac Vien (1975a), p.91.

2. For a detailed discussion of the impact of the Japanese occupation, see Mau (1977). This includes data on the severe taxation imposed as a result of Japanese and French demands during this period.

3. See Hanoi (1960d) and also Duiker (1975b).

4. Hanoi (1970b). The full programme adopted by the ICP at its founding in February 1930 is in Tran Huy Lieu (1960).

5. See Nguyen Xuan Lai (1976), pp.56-58. On the relationship between the ICP and Comintern see Duiker (1975a and 1981), also Hodgkin (1981), and Rousset (1975 and 1978).

6. Duiker (1975b), pp.275-77 and Tran Phuong (1968), pp.52-54.

7. Mau (1977), p.256.

8. For detailed discussion of the 1944-45 famine see Hammer (1955) and Martin (1948).

9. White (1979), p.30.

10. See also Truong Chinh (1963b).

11. Mau (1977).

12. Tran Phuong (1968), p.58.

13. Mau (1977), p.296.

14. Tran Phuong (1968), pp.58-60.

15. Moise (1977), pp.248-52 and Fall (1969), p.119.

16. Tran Phuong (1968), pp.74-82 and Moise (1977), pp.258-59.

17. For the 1952 tax rates see Fall (1960) citing Cu'u Quoc No. 2080 (Hanoi, July 1952), Tran Nhu Trang (1972) and Nguyen Van Vinh (1961). For additional details see VNA 1/1/51, SWB V No.90, p.63; VNA 9/5/51, SWB V No.108, p.62; VNA 5/6/51, SWB V No.112, p.61; and VNA 1/8/53, SWB V No.276, p.32.

18. Mau (1977).

19. See, for example, VNA 23/9/51, SWB V

No.129, p.67.

20. However, the DRV presented many regional and national production estimates, for which see contemporary reports in SWB V and Le Chau (1966) who presents such figures pp.121 and 136.

21. Mau (1977) discusses what French policy existed during this period.

22. VNA SWB V No.108, pp.63-64. This speech also confirmed the line laid down by the 2nd Congress of the VWP in February 1951 at which the CC report was presented by Truong Chinh. Truong Chinh (1977).

23. Le Chau (1966), p.114.

24. VNA 1/9/50, SWB V No.73, p.60.

25. VNA 5/6/51, SWB V No. 112, p.61.

26. VNA 3/8/51, SWB V No. 120, p.68.

27. VNA 4/6/50, SWB V No.60, p.52.

28. Le Chau (1966), p.205.

29. Tran Phuong (1968), p.87.

30. Le Van Luong, cited by Moise (1977), p.268; see also Turley (1980), p.192.

31. For example, see the report of the cadres' conference convened by the VWP CC in August 1948, Tran Phuong (1968), pp.62-64.

32. White (1979), pp.30-31.

33. For this see Hanoi (1970b). Excerpts from the 1930 Theses also appear in Nguyen Khac Vien (1974), pp.277-82 and (1975a), pp. 233-40.

34. White (1979), p.32.

35. Mau (1977), p.271.

36. See Hanoi (1960d) and Duiker (1975b), pp.275-77.

37. Elliot (1980), p.204.

38. See Tran Phuong (1968), p.68.

39. Ibid., pp.77-82.

40. For a discussion of the dissatisfaction of the poor peasants with the relatively slow pace of socio-political change at village level, see White (1979), in addition to numerous contemporary reports, e.g. VNA 12/5/51, SWB V No.109.

41. Variations in 1953 landholding figures depend largely on how communal land is treated and also on that dispersed by landlords.

42. On this point see Moise (1977), pp.254-56.

43. See Appendix Table I which analyses the data from Hanoi (1960c), opp. p.66 Table B in greater detail.

44. Ibid.

45. Moise (1977), p.257.

46. Tran Phuong (1968), p.77.

47. The absence of a single centralised administration in the DRV area from 1945-54 means that no continuous production series is available; these figures are derived from data in a number of sources including the following: Le Chau (1966), pp.121 and 136; Vo Nhan Tri (1967), p.139; Nguyen Tien Hung (1977), pp.4-5; Karamyshev (1959), p.23; and contemporary reports in SWB V.

48. Nguyen Tien Hung (1977), p.9 and references cited therein, p.17. Gourou (1945), p.551 refers to an average ration for Tonkin peasants of 400 grams of white rice per day, equivalent to 630 grams of unhusked rice (paddy) per day, or 230kg per year. Fall (1956), p.116 gives a much higher annual per capita requirement of 452kg of paddy.

49. Pham Van Dong refers to this problem in a speech reported VNA 23/9/51, SWB V No.129, p.67.

speech reported VNA 23/9/51, <u>SWB V</u> No.129, p.67.

50. Tran Phuong (1968), pp.156 and 160. See also Table 3.5.

51. See Table 3.7 where this is discussed further.

52. See the speech by Truong Chinh reported VNA 8/5/51, <u>SWB V</u> No.108, p.64.

Notes to Chapter III

1. Tran Phuong (1968), p.112.

2. See Table 3.1

3. H 13/3/53, <u>SWB V</u> No.235, p.43.

4. VNA 16/3/53, <u>SWB V</u> No.237, pp.55-57.

5. Ho Chi Minh (1962-64) Vol. III, p.425.

6. VNA 20/12/53, <u>SWB V</u> No.316, p.47.

7. Mao Tse Tung (1971), pp.11-39.

8. For a critique of this by the Viet Minh in July 1953 see <u>SWB V</u> No.267.

9. Data in Hanoi (1960c) Table B, opp. p.60 show an average land availability per family of 0.76 hectares in 1945 and 0.62 hectares in 1953.

10. Incomplete translated versions of the Population Classification Decree of 2 March 1953 appear in Cole (1956), pp.139-47 and Fall (1956), pp.172-78; see also Nguyen Van Vinh (1961), p.82. The April Decrees were published on 20 May and reported by VNA 22/5/53, <u>SWB V</u> No.254, pp.29-31 and Voice of Vietnam 28-31 May, <u>SWB V</u> No.259, pp.55-59.

11. White (1979), p.33 and Le Chau (1966), p.133.

12. VNA 26/12/52, SWB V No.213, p.27.

13. The following discussion of this is based on Tran Phuong (1968) especially pp.99, 115 and 125, VNA 28/8/53, SWB V No.282, p.31, and Voice of Vietnam 22/9/53, SWB V No.291, p.56. It is also discussed by Fall (1956), p.132 and Le Chau (1966), p.134 who date it incorrectly to September when it was made public.

14. Tran Phuong (1968), pp.125-26.

15. Ibid. p.158.

16. White (1979), p.34 citing Nhan Dan 6-10/10/53.

17. The information in the following paragraphs is based on Tran Phuong (1968), pp.116-20 and Moise (1977). For accounts of Rent Reduction and Land Reform in individual villages see, for example, Pham Cuong & Nguyen Van Ba (1976), Chaliand (1968), and Burchett (1957). For comparison, the best account of Land Reform in a Chinese village in Hinton (1968).

18. Tran Phuong (1968), p.99 and VNA 8/1/54, SWB V No.320, p.57.

19. Hanoi (1955b).

20. Tran Phuong (1968), p.100, Moise (1977) and contemporary reports in SWB V Nos.311-16, December 1953.

21. Truong Chinh (1977), pp.467-555.

22. Fall (1956), p.126 citing contemporary Vietnamese sources.

23. For the December law see Hanoi (1955b) and for a version of the April Decrees Voice of Vietnam 28-31 May, SWB V No.259, pp.55-59.

24. See Le Chau (1966), pp.122-35 who discusses the Land Reform laws and amendments and

presents the texts, pp.393-406. Moise (1977) discusses the amendments, pp.316-20. Tran Phuong (1968) treats the decrees, laws, and amendments as a single composite law.

25. For a discussion of 'class-ism' during this period see Moise (1980).

26. Tran Phuong (1968), pp.103-4.

27. The details here are taken largely from Tran Phuong (1968), pp.146-62 and Le Chau (1966), pp.121-24. See also the versions of the Population Classification Decree in Cole (1956), pp.139-47 and Fall (1956), pp.172-78.

28. Tran Phuong (1968), p.151--data from 99 villages in four provinces.

29. Articles 2-4 of the Agrarian Reform Law, Hanoi (1955b).

30. Data from this sample are analysed further in Appendix Table I. The comments made in Chapter II regarding the agrarian reform data apply equally in this section.

31. Tran Phuong (1968), p.157.

32. Communique of the 10th Plenum of the 2nd VWP CC reported VNA 29/10/56, SWB V No.610, p.40.

33. Tran Phuong (1968), p.131.

34. Moise (1977), p.419 citing a Soviet source.

35. Tran Phuong (1968), p.125.

36. For studies along these lines see White (1978 and 1979), Moise (1976, 1977, and 1980), Le Chau (1966), Vo Nhan Tri (1967), Tran Phuong (1965 and 1968), and Mau (1977).

37. White (1978), pp.9-10 and Tran Phuong (1968), p.119.

38. Moise (1980), p.93.

39. Tran Phuong (1968), p.189 translated in Moise (1980), p.97. Tran Phuong, pp.193-94 and 199 refers to the class origins of the VWP members.

40. See White (1979), pp.41-42 for a discussion of this.

41. Elliot (1980), p.204. The political struggle and its impact on the VWP is examined in detail in the references cited in footnote 36 above.

42. Burchett (1957).

43. H 26/7/56, SWB V No.583, p.41.

44. H 30/11/56, SWB V No.619, p.23.

45. This letter published in Nhan Dan 18/8/56 is in Ho Chi Minh (1967), pp.304-6. Ho had discussed Khrushchev's speech at the 9th Plenum of the 2nd VWP CC in April 1956. For this see Ho Chi Minh (1962-64), Vol. IV.

46. Communique of the 10th Plenum of the 2nd VWP CC reported, VNA 29/10/56, SWB V No.610, p.41.

47. Ibid., pp.40-41.

48. The September Plenum of the 2nd VWP CC was followed by meetings of senior VWP cadres, 8-21 October; senior army cadres, 23-31 October; the Fatherland Front Central Committee, addressed by Truong Chinh, on 31 October; and the Council of Ministers. Giap's speech was made on 29 October, and Ho Chi Minh also addressed a major meeting on 30 October. For these, see contemporary reports in SWB V Nos.610-14, November 1956. Lengthy excerpts from the Giap speech appear in Nguyen Khac Vien (1970), pp.90-94.

49. The decisions of the Council of Ministers were reported VNA 2/11/56, SWB V No.611, p.32 and VWP changes VNA 29/10/56, SWB V No.610, p.39.

50. The classification amendments were reported H 11/11/56, SWB V No.614, pp.30-32.

51. For detailed accounts of Rectification, see Tran Phuong (1968), pp.188-217, Moise (1977), pp.378-427, Le Chau (1966), pp.141-57 and Nguyen Khac Vien (1970).

52. 'When the elimination of the village notables failed to yield enough land for the hungry landless, rich and middle peasants, among whom included VWP members and cadres, suffered from random confiscation which could not be checked.' Mau (1977), p.318.

53. 'It could be argued that middle-peasants had been the revolutionary vanguard in the anti-colonial or national revolutions.' White (1978), p.7.

54. Moise (1980), p.98.

55. For details of this see VNA 16/11/56, SWB V No.615, p.28, H 20/11/56, SWB V No.616, p.23, VNA 25/11/56, SWB V No.617, p.25, and Le Chau (1966), pp.151-53. The alleged ringleaders of these uprisings subsequently received 15-year prison sentences. VNA 30/4/57, SWB V No.661, pp.42-43.

56. H 9/11/56 SWB V No.614, p.32.

57. Tran Phuong (1968), p.160.

58. The discrepancy in respect of 'rich' peasants is also noted by Gordon (1981), p.41.

59. The accounts of Land Reform in individual villages, referred to in footnote 17 above, give local data. Moise (1977), pp.335-38 and Limbourg (1956), p.70, cite landholding data for particular villages, districts, and provinces.

60. From 1945 to Land Reform, which occurred in 1956 in the vast majority of villages. Some redistributions of French-owned land in particular may even have occurred as far back as 1940.

61. For example, Nguyen Xuan Lai (1957), p.36 gives figures of 2,116 square metres prior to Land Reform and 2,159 square metres post-Land Reform.

62. Truong Chinh cited by Le Chau (1966), pp.155 and 156.

63. Tran Phuong (1968), p.177.

64. White (1978), p.43 discusses the question of reconciliation within VWP branches.

65. Ibid. p.41 refers to difficulties arising from the tendency of the poor 'to have a dichotomous and graded class concepts.' Moise (1980) characterises the process as 'class-ism'--the notion that class status determines everything.

66. It is the former role that is normally emphasised, for example, Le Chau (1966), Moise (1977 and 1980), and White (1979).

67. For a more sophisticated approach to peasant classification, see Patnaik (1976) and for a critique of this and other 'snapshot' classifications, see Shanin (1980).

Notes to Chapter IV

1. Figure 1 shows paddy and food production over the period 1954-75.

2. It was less than one percent in certain provinces, Elliot (1975), p.171.

3. Nguyen Xuan Lai (1957), p.33. North-south trade prior to partition is detailed in Nguyen Tien Hung (1977), pp.3-35.

4. Le Thanh Nghi in Nhan Dan 4/3/74, cited SWB WS No.768, p.A25.

5. Individual households retained family plots within the cooperative equal to 5% of the total area. In addition, there was a varying degree of *de facto* privatisation of plots and production. In recent years this has become increasingly *de jure*.

6. Discussed at length in Elliot (1975 and 1976).

7. Official sources cited by Le Chau (1966), p.265 and Vo Nhan Tri (1967), pp. 260 and 358; see also his discussion of the supply/demand problems, pp.247-48.

8. Referred to in the Report of the Finance Minister to the National Assembly, H 12/9/57, SWB ES No.285, p.32.

9. See, for example, H 6/11/56, SWB ES No.241, p.15.

10. The publication of this decree which prescribed severe penalties, was accompanied by recognition that price destabilisation had affected the National Plan, H 26/5/57, SWB V No.668, p.47.

11. See, for example, VNA 10/10/57, SWB ES No.291, p.29.

12. Le Chau (1966), p.209.

13. In the years from 1959 to 1975 no 10th month harvest exceeded three million tons. Hanoi (1977c), p.120.

14. H 26/6/58, SWB V No. 781, p.49.

15. Doan Trang Truyen and Pham Thanh Vinh (1964), p.139.

16. The stratification of the peasantry is discussed further below.

17. VNA 21/5/51, SWB V No.110, p.72.

18. VNA 11/8/55, SWB ES No.178, p.18. Land Reform had been completed in 735 villages at this time.

19. Appendix Table III details the development of MATs and cooperatives in numerical and percentage terms.

20. Report by Peasants' National Liaison Committee, VNA 29/1/58, SWB ES No.303, p.37.

21. VNA 7/3/56, SWB ES No.206, p.28.

22. VNA 3/7/56, SWB ES No.223, p.36.

23. Le Chau (1966), p.206 and H 29/1/58, SWB ES No.304, p.24.

24. VNA 2/10/57, SWB ES No.287, p.26.

25. National Model Peasants' Conference VNA 30/5/57, SWB ES No.269, p.24.

26. For the decree see, H 10/12/57, SWB ES No.297, p.27.

27. H 31/1/58, SWB ES No.304, pp.24-26.

28. VNA 11/10/58, SWB V No.811, p.46.

29. VNA 7/10/58, SWB ES No.339, p.26.

30. Truong Chinh (1959), p.14.

31. For lengthy extracts from the text of the decisions of the Council of Ministers at this meeting, see VNA 24/10/58, SWB ES No.342, pp.29-32.

32. Mao Tse Tung's article 'On the correct handling of contradictions among the people' had been broadcast in June 1957; his report 'On the Question of Agricultural Cooperation' in January 1956, SWB ES No.199, p.34. For these, see Mao Tse Tung (1971), pp.432-79 and pp.389-420.

33. See Elliot (1976), esp. p.456. The Nhan Dan articles were written by Ho Chi Minh under the pseudonym Tran Luc. This was revealed in Nhan Dan 17/5/80, see Post (forthcoming) who discusses the Chinese influence in some detail.

34. VNA 10/1/59, SWB ES No.352, p.63. See also the excerpts from the 14th Plenum of the 2nd VWP CC resolution in Nhan Dan 12/12/60 cited by Elliot (1976), p.460.

35. For these and other unattainable targets see the Council of Ministers' text on agriculture,

VNA 24/10/58, SWB ES No.342, p.29; also Hanoi (1959) and Nguyen Duy Trinh (1958).

36. Extracts from the resolution of the 16th Plenum of the 2nd VWP CC appear in Hanoi (1964c). The Truong Chinh speech was published, Truong Chinh (1959). In addition, they are extensively referred to in contemporary publications.

37. Editorial in Hoc Tap broadcast by VNA 24-25/12/58, SWB ES No.350, pp.79-85--this quotation, p.81.

38. Truong Chinh (1959), pp.43-44, citing the resolution of the 14th Plenum of the 2nd VWP CC. Vo Nhan Tri (1960), p.363 adduces support from Marx for this contention without making clear at what level of the forces of production it might be appropriate. See also the Hoc Tap editorial cited above, footnote 37. For a direct reference to the 'beneficial influence' of the Soviet Kolkhoz movement, see Truong Van Ba (1962), p.117.

39. Le Chau (1966), p.169.

40. Ibid., pp.179-81.

41. Doan Trang Truyen & Pham Thanh Vinh (1964), pp.68-69.

42. Rene Dumont (1965a, p.44. Emphasis in original.

43. Doan Trang Truyen & Pham Thanh Vinh (1964), p.67.

44. Truong Chinh (1959), p.11.

45. For presentations of the differentiation argument see Truong Chinh (1959), Vo Nhan Tri (1960 and 1967), Nguyen Chi Thanh (1963d), Doan Trang Truyen & Pham Thanh Vinh (1964), Le Chau (1966), Elliot (1976), and Nguyen Khac Vien et al. (1967).

46. For an interpretation along these lines see Elliot (1975), p.178.

47. Truong Chinh (1959), p.9.

48. Truong Van Ba (1962), p.118 gives the following average income figures: 1957--124.8; 1958--136.2; 1959--143.3; and 1960--130.5. Le Chau (1966), p.357 gives 1958-60 figures similar to those in Table 4.7 for poor and middle peasants.

49. The failure to recognise explicitly the consequences of socio-economic mobility within the peasantry is a major weakness of existing studies of both the Land Reform and collectivisation periods.

50. Vo Nhan Tri (1967), p.209 discusses the possibilities for 'clandestine usury' as well as speculation. See also Elliot (1976).

51. Truong Chinh (1959), p.10.

52. Ibid., p.11.

53. Nguyen Chi Thanh (1963d), p.234.

54. Ibid., p.234. He does not expand on the 'positive aspects of individual peasant exploitation' or how forcefully the case for continuation of such a system was argued. Truong Chinh (1959), p.33 also refers to the danger of the 'poor' peasantry turning against the VWP.

55. See the quotation above from the June communique of the Council of Ministers.

56. See in particular, Hanoi (1959) and Truong Chinh (1959). These viewpoints are outlined in Nguyen Chi Thanh (1963a), Truong Van Ba (1962), Le Chau (1966), Doan Trang Truyen & Pham Thanh Vinh (1964), and Vo Nhan Tri (1960 and 1967), in addition to numerous policy statements in subsequent years. Tran Nhu Trang (1972) also discusses the role of cooperatives in eliminating the market mechanisms.

57. The conference was reported H 26/9/58, SWB ES No.338, p.53 and H 20/10/58, SWB ES No.341, pp.42-43.

58. Hanoi (1978), p.107.

59. The most explicit recognition that alternatives were presented is in Nguyen Chi Thanh (1963a) and Nguyen Khac Vien et al. (1967). However, all major works cited, including Truong Chinh (1959) implicitly recognise the existence of alternatives, although how seriously they were considered is unclear. The belief that a 'right' and 'left' wing within the leadership argued two different cases--Elliot (1976) and Gordon (1981) among others--is, as always, oversimplified. The wave of optimism and China 'appreciation' appear to have affected all the leadership, including Ho Chi Minh (whose articles and speeches in 1958 may have stimulated it) and Le Duan, as indicated by the quotation cited above.

60. Truong Chinh (1959), p.47. This was laid down by the 14th Plenum of the 2nd VWP CC in November 1958.

61. Reference to the development of cooperatives in advance of VWP policy is made VNA 3/5/49, SWB V No.2, and the decision of a peasant conference to establish cooperatives is reported VNA 2/1/50, SWB V No.38, p.39. An analysis of the history and 'errors' of the early cooperative movement is given by VNA 2/9/50, SWB V No.73, p.60.

62. New China News Agency (NCNA) 12/10/57, SWB ES No.289, p.31. Truong Chinh (1959) also reviews the early development of cooperatives.

63. The number of cooperatives increased as follows: October 1957--33; December 1957--44; April 1958--107; July 1958--133; October 1958--587; December 1958--4,720; January 1959--4,967.

64. For the targets see VNA 12/3/59, SWB ES No.360, p.36 and for the amalgamation details VNA 13/4/59, SWB WS No.1, p.B32. For details of cooperative organisation see Chapter V.

65. Truong Chinh (1959). VNA 19/5/59, SWB WS No.6, pp.B34-35 and NCNA 28/5/59, SWB WS No.7, p.B24.

66. See VNA 14/9/59, SWB WS No.23, p.B31.

67. VNA 15/2/60, <u>SWB WS</u> No.45, p.B18.

68. VNA 16/11/59, <u>SWB WS</u> No.32, p.B40.

69. Truong Chinh (1959), pp.48-55 and Vo Nhan Tri (1960), p.367.

70. <u>Nhan Dan</u> 26/2/59 cited by Elliot (1975), p.179.

71. Truong Chinh (1959), p.48. Unfortunately the ill-defined use of the term 'middle' peasant in contemporary material, including Truong Chinh (1959), is repeated in subsequent analyses such as Elliot (1976) and Vo Nhan Tri (1967). The latter also refers to 'new and old' in the context of 'middle' peasants, (p.277).

72. Le Chau (1966), p.190.

73. Elliot (1975), p.191, footnote 53, cites <u>Nhan Dan</u> on the use of coercion.

74. The ratio of VWP members to population in the DRV was only 2% at the outset of collectivisation and only 40% of members were at the village level. Elliot (1975), p.173. The theme of the peasantry reasserting their autonomy vis-a-vis the State is pursued by Mau (1977).

75. See Truong Chinh (1959), pp.22-24 for discussion of attitudes of various peasant strata towards cooperatives.

76. Elliot (1976), p.471.

77. Doan Trang Truyen & Pham Thanh Vinh (1964), p.72.

78. The failure to recognise openly the size of the 'rich' peasant grouping and the possible reasons for this are discussed by Gordon (1981), p.38-42. The weakness of his analysis is the simple association of a 'right' and 'left' within the VWP leadership with dichotomous 'rich' and 'poor' strata of the peasantry.

79. Vo Nhan Tri (1967), p.208 citing official

DRV sources.

80. Truong Chinh (1959), p.55.

81. H 22/12/59, SWB WS No.38, p.B32.

82. 'There is real benefit in joining the cooperatives--but only for the middle peasants.' Nhan Dan 7/8/59 cited by Elliot (1975), p.178.

83. See Truong Chinh (1959) and contemporary reports. Elliot (1976), pp.472-80 discusses in some detail the effect of the social composition of the VWP during this period.

84. The notion that the leadership was merely responding to pressure from the peasantry in any widespread sense and reversing a previously 'rightist' phase is clearly a misconception, although in ideological terms, it served a purpose as a further rationalisation for the policy adopted.

85. H 30/3/60, SWB WS No.51, p.B33.

86. H 3/7/60, SWB WS No.66, p.B23 and H 30/3/60, SWB WS No.51, p.32.

87. A relatively clear exposition of the need to severely restrict rural living standards in order to build socialism is given in Doan Trang Truyen & Pham Thanh Vinh (1964), pp.138-45.

88. H 18/4/60, SWB WS No.55, p.B30.

89. From an article by Vu Oanh in Hoc Tap, August 1964, cited by Elliot (1975), p.174.

90. H 30/3/60, SWB WS No.51, p.B33.

91. Income of 'middle' peasants in cooperatives was only 91% of that of those working individually. 'Poor' peasant income, however, was 10% higher in than out. Nghien Cuu Kinh Te No.1, 1961, pp.9 and 41, cited by Le Chau (1966), pp.185 and 358.

92. Over the whole period to 1975 it ranks third behind those of 1959 and 1958.

93. VNA 8/9/60, SWB WS No.75, p.B14 reported a speech by Tran Huu Duc.

94. H 30/9/60, SWB WS No.78, p.B13.

95. VNA 30/9/60, SWB WS No.78, p.B13.

96. See Appendix Table III.

97. Truong Chinh (1959), pp.32-33.

98. The data on this are limited but available empirical and non-empirical evidence, as well as the account in Le Chau (1966), accord with the interpretation given here.

99. The same arguments had been used in China, see Walker (1966), and the USSR, see Kahan (1964). 'Among the multitude of arguments in favour of socialised agriculture were (1) achieving economies of scale...(2) changing the terms of trade between the State and the agricultural producers without impairing the agricultural output control over the peasantry, to prevent the breeding of capitalism.' Ibid. p.252.

100. See Walker (1966).

101. Ibid., p.43. It was 'better to be guilty of subjectivist utopian thinking than to be judged an enemy of socialism or of Chairman Mao.'

Notes to Chapter V

1. For the Constitution of Low-Level and High-Level Cooperatives from which quotations are taken, see Hanoi (1964d and 1971). Additional information in this section is from Cuong Pham (1964), Nguyen Xuan Lai, ed. (1977) and official sources cited in Elliot (1976), pp.495-98, and 515-17.

2. Truong Chinh (1959), p.61.

3. See for example, Nguyen Xuan Lai (1967) and Le Chau (1966), pp.356-60.

4. From the resolution of the 16th Plenum of the 2nd VWP CC April 1959, VNA 19/5/59, SWB WS No.6, p.B35.

5. Appendix Table IX gives various estimates of the average number of work days per cooperator.

6. Hanoi (1977c) Table 137, p.180.

7. Nguyen Chi Thanh (1963a), p.232.

8. Dumont (1973), p.89.

9. Nhan Dan 11/11/60, SWB WS No.84, p.B27.

10. Vo Nhan Tri (1967), p.424.

11. Nguyen Chi Thanh (1963a), p.230.

12. VNA 4/12/61, SWB WS No. 139, p.B22.

13. Appendix Table V.

14. Nguyen Chi Thanh (1963a), p.230.

15. The resolution of this meeting was published Hanoi (1964b).

16. For example, H 3/7/60, SWB WS No.66, p.B23.

17. Nguyen Khac Vien et al. (1967), p.85. See also Elliot (1976), p.489. The net departure appears to have been 4.2% of the peasantry between 1961 and 1964.

18. Hanoi (1964b), p.155.

19. Ibid., p.163.

20. Vo Nhan Tri (1967), pp.297-98 also discusses the difficulties of making the collective form a reality.

21. Truong Chinh (1959), p.38.

22. Hanoi (1964b), pp.156-57.

23. Truong Chinh (1959), p.69.

24. Ibid., p.61, Nguyen Xuan Lai (1967), p.100 and Charrière (1966), p.34.

25. Appendix Table VII.

26. For example, Truong Chinh reported by VNA 8/1/60, SWB WS No.40, p.B33.

27. Le Duan (1965) vol. II, p.70. See also Hanoi (1964b).

28. For example, Hanoi (1970a) and Gordon (1981).

29. VNA 15 and 16/5/61, SWB WS No.110, pp.B32-33 and VNA 14/6/61, SWB WS No.115, p.B28.

30. VWP CC instructions issued 31/12/61, H 9/1/62, SWB WS No.147, p.B28.

31. Deputy Premier Pham Hung at an agricultural conference in Hanoi H 10/3/62, SWB WS No.153, p.B18.

32. H 25/11/61, SWB WS No.139, pp.B21-22.

33. VNA 6/12/60, SWB WS No.87, p.B23.

34. Deputy Premier Pham Hung in H 10/3/62, SWB WS No.153, p.B18.

35. Ibid.

36. Ibid., p.B19.

37. Ibid.

38. Luu Dong & Tran Chau, 'An Initial Effort in the Leadership Over Agricultural Production in Bac Ninh Province' H 27/3/62, SWB WS No.156, pp.B16-18.

39. Ibid., p.B16.

40. Hanoi (1964b), pp.157 and 159.

41. Truong Chinh (1959), p.55.

42. These tendencies were also noted by the 5th Plenum, Hanoi (1964b), p.155 which remarked on 'fairly prevalent bureaucracy,' p.156.

43. Data from Hanoi (1977c), pp.118-20.

44. From broadcast text of a talk by the Deputy Agriculture Minister H 3/4/62, SWB WS No.156, pp.B19-21.

45. Lavallée (1971), p.53.

46. VNA 25/4/62, SWB WS No.159, pp.B14-15.

47. Ibid.

48. Dumont (1965a), p.42.

49. Nhan Dan 28/7/59 cited by Elliot (1976), p.456.

50. Nguyen Xuan Lai (1971), p.125.

51. Deputy Premier Nguyen Duy Trinh H 19/7/62, SWB WS No.172, p.B11.

52. H 6/8/62, SWB WS No.174, pp.B25-26, from which the following quotations are also taken.

53. Ibid., p.B26.

54. H 25/9/62, SWB WS No.181, p.B13.

55. H 19/10/62, SWB WS No.185, p.B13.

56. Elliot (1976), p.448.

57. Le Duan (1965), vol.III, p.18 (emphasis added).

58. For details of this policy see H 4/7/62, SWB WS No.170, p.B19.

59. Tran Nhu Trang (1972) and Le Chau (1966), p.281.

60. Charrière (1966), p.32.

61. Ibid., p.40, and Nguyen Xuan Lai (1967). Elliot (1976), p.499 refers to the tax averaging 7.2% of total food production and 4.8% of total agricultural production for the period 1960-62.

62. Vo Nhan Tri (1967), p.383. See also Table XI and Figs.2-7.

63. Ibid., p.452 and H 4/7/62, SWB WS No.170, p.B19.

64. H 4/7/62, SWB WS No.170, p.B19.

65. Appendix Table III.

66. Le Duan (1965) vol.II, p.97.

67. H 10/11/62, SWB WS No.188, p.B17.

68. For details of the official policy regarding distribution of collective income see Hanoi (1964d and 1971); also Elliot (1976), pp.495-98.

69. H 8/12/62, SWB WS No.192, p.B27.

70. H 10/12/62, SWB WS No.192, p.B30.

71. The impact of droughts and floods for 1955-63 is detailed in Lavalée (1971), p.53.

72. Appendix Table VI.

73. For summary data on cooperatives see Appendix Table III.

74. Elliot (1976), pp.515-17.

75. See for example, the report by Luu Dong and Tran Chau, 'An Initial Effort in the Leadership Over Agricultural Production in Bac Ninh Province,' H 27/3/62, SWB WS No.156, pp.B16-18.

76. H 8/3/63, SWB WS No.203, p.B27. This campaign is extensively discussed in numerous contemporary broadcasts and publications. See also Vo Nhan Tri (1967), pp.415-16, Elliot (1976),

pp.511-13 and Nguyen Khac Vien et al. (1967).

77. Vo Nhan Tri (1967), p.416.

78. Ibid., pp.416-17 and Elliot (1976), pp.512-13.

79. For 1959 figures see Table 4.6. The 1963 figures are from Nhan Dan 30/4/63 cited by Elliot (1976), p.494.

80. Appendix Table III.

81. Hanoi (1977c) Table 137, p.130.

82. For a summary of this resolution, see VNA 28/4/63, SWB WS No.211, pp.B14-21. See also Elliot (1975), p.181.

83. H 26/4/63, SWB WS No. 213, p.B28.

84. Ibid.

85. Ibid., p.B29.

86. Elliot (1976), p.499.

87. Dumont (1965b), p.19.

88. Elliot (1976), p.499 and H 26/4/63, SWB WS No.213, p.B29.

89. H 4/11/65, SWB WS No.341, p.A43.

90. H 26/4/63, SWB WS No.213, p.B30.

91. Nguyen Xuan Lai (1971), p.150.

92. VNA 5/1/64, SWB WS No.246, p.B35 and other contemporary reports.

93. Nguyen Tien Hung (1977), p.66.

94. Hanoi (1977c), Table 102, p.143. Vo Nhan Tri (1967), p.453 refers to 'errors' by State commerce in 1963.

95. Nguyen Tien Hung (1977), pp.121-23 and H 15/11/63, SWB WS No. 241, p.B34.

96. Increasing land utilisation is discussed in an address by Bui Cong Trung reported VNA 1/10/62, SWB WS No.182, p.B10 and VNA 22/11/62, SWB WS No.190, p.B20.

97. VNA 21/5/58, SWB ES No.319, p.41.

98. The deficiencies of highland development in this respect are commented upon in Nhan Dan 21/1/63, SWB WS No.198, p.B20.

99. It was 114,000 tons in 1960, Le Chau (1966), p.198.

100. Nhan Dan 17/10/60, SWB WS No.80, p.B27.

101. Contemporary reports and Nguyen Chi Thanh (1963a), p.236.

102. For details see Tran Quan Thuyet and Dao The Thanh (1974), VNA 5/11/61, SWB WS No.139, p.B22, and VNA 5/7/65, SWB WS No.323, p.A39.

103. VNA 25/4/62, SWB WS No.159, p.B15 and VNA 22/11/62, SWB WS No.190, p.B20.

104. See for example, VNA 9/6/62, SWB WS No.166, p.B19.

105. For details of the State farms and their role, see for example, H 19/2/65, SWB WS No.305, reporting Premier Pham Van Dong.

106. Nguyen Xuan Lai (1967).

107. VNA 13/11/61, SWB WS No.136, p.B19.

108. From VWP CC 'Instructions on stepping up the movement for improving farm tools and the initial mechanisation of agriculture' cited in H 5/1/64, SWB WS No.246, p.B32.

109. Charrière (1966), pp.34-37, Nguyen Xuan Lai (1971), pp.142-43 and Nguyen Tien Hung (1977), p.150.

110. Nguyen Xuan Lai (1971), p.165.

111. VWP CC 'Instructions on stepping up...', H 5/1/64, SWB WS No.246, pp.B30-33.

112. White (1982a), p.14 refers to the 'popularisation of low-level appropriate technology.'

113. For data on labour productivity see Appendix Table VIII.

114. Le Duan (1965), vol.II, p.22.

115. For a presentation of this view see Gordon (1981), esp. pp.30-35.

116. The different views in the on-going intra-VWP debate are discussed in Tran Nhu Trang (1972).

117. Le Duan (1965), vol.II.

118. H 30/5/64, SWB WS No.269, p.B29.

119. See for example H 30/5/64, SWB WS No.269, p.B30.

120. H 10/6/64, SWB WS No.269, p.B32.

121. The inadequacy of industrial production for agriculture is detailed in Nguyen Xuan Lai (1971), pp.142-43.

122. Elliot (1976), pp.491-92. Le Chau (1966), p.188 gives a figure of 3%.

123. Truong Chinh (1959), p.74. Peasant antipathy towards cooperatives is discussed in Dumont (1965a and 1965b), pp.18-24.

124. Appendix Table VI.

125. Nguyen Duy Trinh VNA 1/1/65, SWB WS No.297, p.B26.

126. H 8/6/64, SWB WS No.269, p.B33.

127. Doan Trang Truyen and Pham Thanh Vinh (1964) p.86.

128. Lavallée (1971) p.57. In 1965 they were

23.5% of the total, Nguyen Xuan Lai and Vu Quoc Tuan (1976) p.234. Although no details are available on the composition of consumer goods imports, food imports were insignificant at this time.

129. See Vo Nhan Tri (1967), pp.411-15; see also VNA 31/12/63, SWB WS No.246, p.B35.

130. VNA 26/5/64, SWB WS No.271, p.B23.

131. H 3/6/64, SWB WS No.268, p.B25.

132. VNA 9/10/61, SWB WS No.131, p.B23 and VNA 25/1/65, SWB WS No.300, p.B27.

133. VNA 14/3/66, SWB WS No.358, p.A41.

134. Nguyen Chi Thanh reported H 10/6/64, SWB WS No.269, p.B32.

135. H 29/6/64, SWB WS No.272, p.B20.

136. Truong Chinh (1959), pp.84-98.

137. Elliot (1976), p.510. The information on the VWP in the following paragraph derives from this source, pp.504-11.

138. Pham Hung in a speech criticising cadres who favoured the family economy.

139. VNA 19/1/65 SWB WS No.299, pp.B26-33.

140. Ibid., p.B30.

141. Ibid., p.B29. See also Vo Nhan Tri (1967), p.512.

142. H 6 and 7/10/65, SWB WS No.337, pp.A40-50.

143. Ibid., p.A42.

144. Ibid., p.A46.

145. Ibid., p.A42. Nguyen Xuan Lai (1967) gives a target of 220 to the cooperative and 25 to the State; Elliot (1976), pp.497-98 cites a target of 200 to the cooperative, 50 to the State in

labour service, and 50 to the family plot.

146. Elliot (1976), p.490 and Le Duan (1965), p.197.

147. Charrière (1966), p.40.

148. Elliot (1976), p.456.

149. Nguyen Xuan Lai (1971), p.170 and Dumont (1973), p.89.

150. Hanoi (1977c) Table 83, p.122; see also Appendix Table IV.

151. Elliot (1976), p.494 citing detailed enquiries published in Nhan Dan 28/4/64.

152. Hanoi (1977c), Table 137, p.120.

153. Elliot (1976), pp.493-94.

154. Hanoi (1964b), p.157. This was repeated at the 8th Plenum in April 1963, VNA 28/4/63, SWB WS No.211, p.B21.

155. Elliot (1976), p.494.

156. Hanoi (1977c), Table 137, p.180 and Table 72, p.111.

157. Appendix Table IV.

158. Nhan Dan 4/11/65, SWB WS No.341, pp.A42-44.

159. The data in this paragraph are presented in Appendix Table III.

160. Elliot (1975), p.184.

161. Perkins (1969), Ishikawa (1971), among others. Nolan (1971), p.217 argues that there were substantial returns at least in the 1950s in China. He also argues that only collectives could provide access to these returns. For evidence that private agricultural production can also access them, see Hsieh (1971).

162. For the role of private agriculture in the USSR, see Hill (1975), Lewin (1980), Wadekin (1971) and (1973), and Kerblay (1980). For China, see Walker (1965). Lewin (1980) makes many of the same points regarding the USSR that are made here concerning the DRV.

Notes to Chapter VI

1. H 26/12/65, SWB WS No.349, pp.A56-62 (emphasis added).

2. Nhan Dan 17/2/66, SWB WS No.355, pp.A34-37.

3. H 9/11/66, SWB WS No.393, p.A13.

4. H 26/1/66, SWB WS No.352, p.A49.

5. Hanoi (1977c) Table 104, p.145.

6. Ibid. Table 136, p.179.

7. Ibid. Tables 132 and 136, pp.175 and 179. See also Dumesnil (1970) p.545 and Nguyen Xuan Lai and Vu Quoc Tuan (1976), p.231.

8. H 27/3/66, SWB WS No.362, pp.A46-47. This is a report of a conference convened by the State Price Commission.

9. Appendix Table VI presents available data on State procurement and sales.

10. Burchett (1966).

11. Nguyen Xuan Lai and Vu Quoc Tuan (1976).

12. Pike (1972) and Van Dyke (1972).

13. See Appendix Table III for summary data on collectivisation.

14. Van Dyke (1972), p.178 gives data for the number of families and land for 1966 for different

sub-regions.

15. Nhan Dan 17/2/66, SWB WS No.355, pp.A34-37.

16. Nguyen Con to the 1966 National Assembly, H 27/4/66, SWB WS No.366, pp.A24-41. See also Appendix Table XI.

17. Van Dyke (1972), p.192. Tran Nhu Trang (1972) also discusses the shift to local industry in some detail.

18. Nguyen Xuan Lai (1971), p.165 and Van Dyke (1972), p.193.

19. Van Dyke (1972), p.193.

20. These points are taken from the major summary of the development of the economy given by Deputy Premier Pham Hung at a conference in early 1967 to discuss the implementation of the 1966 and 1967 State plans which had replaced the proposed 1966-70 Second Five-Year Plan. H 4 and 5/2/67, SWB WS Nos. 406, pp.A6-29 and 407, pp.A5-7. For a further discussion of local industry see VNA 5/12/67, SWB WS No.447, p.All.

21. Resolution of the 3rd VWP Congress in 1960, quoted by Le Thanh Nghi, Nhan Dan 4/11/67, SWB WS No.450, p.A20.

22. See Appendix Table XI.

23. Nguyen Xuan Lai and Vu Quoc Tuan (1976).

24. For example, VNA 22/8/66, SWB WS No.381, p.A14.

25. Appendix Table V.

26. VNA 4/2/66, SWB WS No.354, p.A37.

27. Pham Van Dong's report to the National Assembly 16/4/66, reported VNA 28/4/66, SWB III No.2166, p.C2.

28. Nhan Dan 22/11/66, SWB WS No.395, p.A15.

29. H 17/6/70, SWB WS No.577, p.A13. See also Nhan Dan 3/8/67, SWB WS No.430, p.A8.

30. Appendix Table VIII.

31. For some criticisms of heavy industry performance, see Dumesnil (1970), p.535 and Charrière (1966), p.36.

32. H 4/2/67, SWB WS No.406, p.A17.

33. Ibid., p.A19.

34. Nhan Dan 9/8/67, SWB WS No.432, pp.A7-9.

35. Nhan Dan 15/11/67, SWB WS No.445, pp.8-9.

36. H 29/5/67, SWB WS No.423, p.A10. Livestock data are presented in Appendix Table V.

37. Van Dyke (1972), pp.172-74.

38. The Council of Ministers' resolution on the local economy, broadcast H 21 and 22/10/68, SWB WS Nos. 492, pp.A22-26, No.493, pp.20-25, and No.494, pp.19-21. This foreshadowed future emphasis on the district as the 'appropriate' unit of rural organisation.

39. See Nguyen Con's report to the National Assembly, H 25/5/68, SWB WS No.471, pp.A18-22.

40. Tran Nhu Trang (1972) cites numerous examples and discusses 'petty corruption.'

41. Nhan Dan 26/5/69, SWB WS No.522, p.A12. Numerous other quotations of this kind can be found in contemporary reports; see also Van Dyke (1972), pp.167-69.

42. White (1982b), p.13, also Van Dyke (1972), p.179.

43. See Trang (1972), Van Dyke (1972), p.179, and Turley (1972b) as well as innumerable contemporary reports.

44. See also White (1982b), and for a detailed exposition of this theme, Elliot (1975) and Mau

(1977).

45. See the speech by Truong Chinh to Vinh Phu province cadres--6 November 1968--published *Nhan Dan* 29 and 30/1/69 and in translation by US sources; Truong Chinh (1970); and also a series of *Nhan Dan* articles and editorials, December 1968 and January 1969.

46. The relative weakness of the degree of control over all aspects of economic management and its deterioration during the war years was referred to repeatedly during this period. See, for example, *Nhan Dan* 2/1/69, *SWB WS* No.503, pp.A24-25.

47. *Nhan Dan* 11/12/68, *SWB WS* No.499, p.A22.

48. *Nhan Dan* 10/1/69, *SWB WS* No.504, pp.A37-38.

49. Ibid., p.A37. See also Truong Chinh (1970) and Gordon (1982).

50. Truong Chinh (1970).

51. Truong Chinh (1959).

52. Truong Chinh (1970).

53. For example, Nguyen Xuan Lai (1967).

54. H 8/8/69, *SWB WS* No.532, pp.C1-5; also Hanoi (1971). For a summary of cooperative regulations see Chapter V.

55. Turley (1972b), pp.30-31.

56. H 26/10/69, *SWB WS* No.544, p.A18.

57. Nguyen Xuan Lai (1971), p.153.

58. *Nhan Dan* 12/1/69, *SWB WS* No.505, p.A20.

59. Gordon (1982) and Trang (1972).

60. Dumesnil (1970), p.545.

61. H 3/10/74, *SWB WS* No.799, pp.C1-14.

62. Van Dyke (1972), p.163 shows the distribution of workdays throughout the year.

63. Elliot (1976), p.452.

64. Ibid., p.491.

65. H 7/1/70, SWB WS No.554, p.A14. This referred to factories and cooperatives.

66. Nhan Dan 5/6/70, SWB WS No.575, p.A16.

67. Nhan Dan 18/12/71, SWB WS No.612, p.A11.

68. Nguyen Xuan Lai (1971), Dumont (1965a), and Elliot (1976), p.456.

69. Nhan Dan 12/1/72, SWB WS No.658, p.A25; also see Appendix Table IX.

70. Nguyen Xuan Lai (1967) gives a breakdown of family income as follows: animal husbandry 36%, crops 24%, handicraft 18%, others 22%.

71. See, for example, H 21/3/69, SWB WS No.513, p.A15 and for similar points related to the construction industry, H 17/2/70, SWB WS No.560, p.A11.

72. Hanoi (1963b), p.62.

73. See, for example, the Resolution of the Council of Ministers on 'Productive Labour,' H 12/2/70, SWB III No.3306, p.B2.

74. H 15/5/70, SWB WS No.571, pp.C1-5. A lengthy speech by Le Thanh Nghi on this resolution was broadcast the following day H 16/5/70, SWB WS No.572, pp.C1/1-6.

75. H 16/5/70, SWB WS No.572, p.C1/1.

76. Ibid., p.C1/2.

77. H 15/5/70, SWB WS No.571, p.C2.

78. H 16/5/70, SWB WS No.572, p.C1/2. The State probably couldn't even if it had wanted to.

79. H 15/5/70, <u>SWB WS</u> No.571, p.C3. For further details of the distribution system and certain changes over time see Van Dyke (1972), pp.180-82.

80. H 16/5/70, <u>SWB WS</u> No.572, pp.Cl/2.

81. <u>Nhan Dan</u> 24/12/70, <u>SWB III</u> No.3571, pp.B1-2.

82. H 16/5/70, <u>SWB WS</u> No.572, pp.Cl/3.

83. Hanoi (1977c) Table 104, p.145.

84. Nguyen Xuan Lai and Vu Quoc Tuan (1976), p.231 show a fall in free market prices between 1968 and 1972.

85. 'Imbalances' and 'irrationalities' in urban-rural trade and the need for rapid price adjustment are referred to in <u>Nhan Dan</u> 3/7/70, <u>SWB WS</u> No.572. p.A18.

86. Van Dyke (1972), p.180.

87. Burchett (1966).

88. H 16/5/70, <u>SWB WS</u> No.572, p.Cl/4.

89. See Nguyen Con's report to the June 1970 National Assembly meeting, <u>SWB III</u> Nos. 3402, 3407 and 3408.

90. Broadcast H 2/6/70, <u>SWB WS</u> No.575, pp.Cl/1-8, although it had been issued earlier.

91. Ibid., p.Cl/2.

92. Ibid. p.Cl/6.

93. There are innumerable contemporary references to this, including criticism of the use of cooperative grain to feed private pigs. See also Van Dyke (1972), p.179.

94. Pham Van Dong in his report to the 7th Session of the National Assembly which met 2-4 March 1971, VNA 5/3/71, <u>SWB III</u> No.3628, p.B7.

95. See H 3/12/71, SWB WS No.651, p.A24 and other contemporary reports. Precise data on food imports are unavailable.

96. For information on this crop and its development see: Dao The Tuan (1974), pp.51-72; Nguyen Ngoc Luu (1982), pp.72-77; Gordon (1974), and Duong Hong Hien (1974). For an earlier discussion of DRV attempts to develop rice strains see Bui Huy Dap (1967).

97. Hanoi (1977c) Tables 82-84, pp.121-23.

98. Van Dyke (1972), p.170.

99. H 23/9/70, SWB III No.3492, p.B2. This showed the problem of supervising and assessing work.

100. Nhan Dan 12/1/72, SWB WS No.658, p.A25.

101. H 20/6/72, SWB WS No.679, p.A19.

102. Nhan Dan 12/9/72, SWB WS No.691, pp.A25-27.

103. For comprehensive data on industrial crops which accounted for 4-5% of the total cultivated area, see Hanoi (1977c), Tables 86-91, pp.125-30.

104. Dumesnil (1970), p.538 cites a meat ration figure of 3.6kg per year. See also Nguyen Xuan Lai and Vu Quoc Tuan (1976), p.232.

105. Nhan Dan 5/5/73, SWB WS No.724, p.All.

106. Ibid.

107. Ibid.

108. Nhan Dan 8/5/73, SWB WS No.725, p.A19. This report cited instances of cracked, leaking, and subsiding irrigation stations. Data in Hanoi (1977c) Table 75, p.114 show effective utilisation at only 67% of capacity.

109. Nguyen Tien Hung (1977), p.124 and data

on domestic production, p.144.

110. Nhan Dan 1/6/73, SWB WS No.728, pp.A22-23; for discussion of the technical relationships and methods of fertiliser preparation and use, H 27/12/66, SWB WS No.399, p.A19.

111. Van Dyke (1972), p.187.

112. Tran Quan Thuyet and Dao The Tuan (1974); Luu (1982), pp.80-84.

113. Hanoi (1977c).

114. For two articles on seeds during this period, see Nhan Dan 23/5/73, SWB WS No.727, pp.A19-20, and Nhan Dan 28/5/73, SWB WS No.728, pp.A22-23.

115. See for example Buu Huy Dap (1971), and White (1982a), pp.15-16.

116. See Luu (1982) for a presentation which appears to totally ignore the actual performance of DRV agriculture and the limited or non-implementation of the technical developments described on pp.156-64.

117. VNA 1/6/73, SWB WS No.728, pp.A24-25. For data on agricultural mechanisation see Hanoi (1977c), Table 76, p.155, Hanoi (1978) Tables 44-48, pp.100-102. See also Nguyen Tien Hung (1977), p.124, and Van Dyke (1972), pp.166-69.

118. Nguyen Xuan Lai (1971) reviews the priorities for agricultural modernisation.

119. Hanoi (1977c) Tables 10-11, pp.43-44 and Table 73, p.112.

120. From Ministry of Agriculture Crop Directives, H 18/10/65, SWB WS No.339, p.A46.

121. See Appendix Table VIII.

122. H 15/7/73, SWB WS No.735, p.A25.

123. H 15/10/73, SWB WS No.748, p.A22 (emphasis added).

124. *Nhan Dan* 7/9/73, *SWB WS* No.743, p.A22.

125. See for example H 15/10/73 *SWB WS* No.749, p.A23.

126. Appendix Table V.

127. *Nhan Dan* 22/10/73, *SWB WS* No.751, p.A20.

128. Reports of 1973 conference on highland agriculture H 11/8/73, *SWB WS* No.739, p.A30. See also *Nhan Dan* 11/7/72, *SWB WS* No.682, pp.A24-25 and Le Duan (1971).

129. H 20/4/74, *SWB WS* No.775, p.A17.

130. See Le Thanh Nghi's report to the National Assembly 4/2/73, *SWB WS* No.763, pp.A26-33.

131. H 18/2/74, *SWB WS* No.765, pp.A18-19.

132. *Nhan Dan* 3/1/72, *SWB WS* No.656, p.A35.

133. *Nhan Dan* 27/2/74, *SWB WS* No.767, pp.A13-14.

134. *Nhan Dan* 7/2/74, *SWB WS* No.765, p.A21.

135. *Nhan Dan* 27/2/74, *SWB WS* No.767, p.A24.

136. *Nhan Dan* 16/4/74, *SWB WS* No.773, p.A19. For further discussion of these points and examples see *Nhan Dan* 20/2/74, *SWB WS* No.767, pp.A14-15 and *Nhan Dan* 16-19/4/74, *SWB WS* No.773, pp.A18-22.

137. *Nhan Dan* 17/4/74, *SWB WS* No.773, p.A21.

138. See also Appendix Table VIII.

139. See for example *Nhan Dan* 3/5/74, *SWB WS* No.776, p.A24.

140. *Nhan Dan* 19/6/74, *SWB WS* No.783, p.A19.

141. *Nhan Dan* 24/5/74, *SWB WS* No.776, p.A19.

142. For census data see H 29/8/74, <u>SWB WS</u> No.792, p.A25.

143. <u>Nhan Dan</u> 17/6/74, <u>SWB WS</u> No.783, p.A13.

144. <u>Nhan Dan</u> 26/3/74, <u>SWB WS</u> No.769, p.A12.

145. <u>Nhan Dan</u> 5/7/74, <u>SWB WS</u> No.785, pp.A19-20. 'Improvement' should be seen partly as achievement of State objectives.

146. In the following discussion reference is made to the following main documents and reports: (i) Draft Statute on High-Level Agricultural Co-operatives (henceforth Statute) H 3/10/74, <u>SWB WS</u> No.799, pp.C1-14; (ii) VWP CC Directive on the Re-organisation of Agricultural Production (henceforth Directive) H 28/9/74, <u>SWB WS</u> No.796, p.A16-22; (iii) Report by Hoang Anh, Sec'y. of the VWP CC and DRV Vice-Premier to the Conference (Hoang Anh) H 1/10/74, <u>SWB WS</u> No.803, pp.C1-14; (iv) Talk by Le Duan to the Conference (Le Duan) H 2/10/74, <u>SWB WS</u> No.801, pp.C1-11; (v) Talk by Premier Pham Van Dong to the Conference (Pham Van Dong) H 11/10/74, <u>SWB WS</u> No.805, pp.C1-12; (vi) Council of Ministers' Communique on Prices (Prices) H 14/10/74, <u>SWB WS</u> No.798, pp.A30-31. In addition the Conference was extensively reported in <u>Nhan Dan</u> during this period, especially during October. (iii), (iv) and (v) were subsequently published, Le Duan and Pham Van Dong (1975).

147. Directive, p.A16. Between 1964 and 1973 the area of paddy and subsidiary crops fell by 517,000 hectares.

148. Hoang Anh, p.C3.

149. Ibid., p.C4.

150. Ibid.

151. Ibid.

152. Ibid., p.C11.

153. Le Duan, p.C3.

154. For example, none of the eight policies

referred to by Hoang Anh, pp.C13-14, are genuinely 'new,' all having been previously proposed or introduced, albeit in slightly different forms. Similarly, the NMS (see Nguyen Xuan Lai, ed. (1977) contains little that was new.

155. Prices, pp.30-31.

156. Le Duan, p.C1.

157. Hanoi (1977c), Table 139, p.182.

158. Ibid., Tables 109-11, pp.150-51.

159. Ibid., Table 102, p.43.

160. See for example Nhan Dan 4/1/75, SWB WS No.811, p.A18 and Le Thanh Nghi on the 1975 State plan, SWB WS No.815, p.A17.

161. Le Duan, p.C6.

162. Le Duan, pp.C7-8. For other quotations on the tendency for cooperatives to pursue their own interests, see Turley (1972b), p.43, Nhan Dan 5/5/73, SWB WS No.724, p.A11, and Le Duan (1970), p.99.

163. For a stylised presentation see Nguyen Xuan Lai, ed. (1977).

164. Directive, p.A20. See also Le Duan, p.C10.

165. Le Duan, p.C9.

166. Le Duan, p.C4.

167. Nhan Dan 29/10/74, SWB WS No.805, p.C13.

168. For a discussion of localisation of industry in China see Riskin (1971).

169. Kerblay (1980), p.60 discusses the similar attitude of the Soviet ideologists.

170. Walker (1965), p.56. He details the significance of the private sector to both peasants and government, pp.23-56. Walker (1966) also

discusses this dilemma of raising output and tightening government control. Kerblay discusses the persistence of this dependence in the USSR 'until the collective economy is able to take over the function fulfilled by the family farm, the objective reasons for its survival remain. In the past, restrictions imposed on private agricultural activities generally worked out to the disadvantage of the city dwellers.' Kerblay (1980), p.80.

171. For discussions of managerial problems on Soviet Kolkhoz see Wittfogel (1971), and Bradley and Clark (1972).

172. Nguyen Khac Vien in Gough (1982), p.379.

Notes to Conclusion

1. For a summary of both the mechanisation before and after collectivisation arguments, and the Chinese view prior to 1954 and subsequently, see Walker (1966), esp. p.3 and pp.6-8.

2. Ellman (1975), p.849. See also Kerblay (1980).

3. Tang (1968), pp.491-97, Perkins (1969), pp.189-91, and Ishikawa (1967).

4. For example, Ishikawa (1967) and Hsieh (1971).

5. Tran Phuong in _Nghien Cuu Kinh Te_ cited in Mau (1977), p.379.

6. Le Duan and Pham Van Dong (1975), p.55.

7. _Nhan Dan_ 22/10/79, cited in White (1982b), p.17.

8. This summary largely excludes consideration of the southern half of the SRV.

9. Nguyen Khac Vien in Gough (1982), p.379; see, in addition, his comment, p.380.

10. 'The crops in 1977 were a complete

failure because of drought. People did whatever they could to feed themselves. Some even went begging. It was impossible to mobilise the work-force any more.' From an article on a particular district near Haiphong which was one of the first to introduce the 'reforms,' <u>Vietnam Courier</u>, vol.17, No.2, 1981, p.13.

11. Hanoi (1977a). The VWP changed its name at this Congress to the Communist Party of Vietnam (CPV).

12. Chapter V. For the low-level Statute see Hanoi (1964d); the high-level Statute, Hanoi (1971), the Draft 1974 Statute, <u>SWB WS</u> No.799, pp.C1-14, and the NMS, Nguyen Xuan Lai, ed., (1977).

13. For an outline of the 'product-contract' system see the report of the Ministry of Agriculture Review Conference in August 1981, <u>Nhan Dan</u> 26, 27 and 28 August, and 8, 9 and 10 September 1981, <u>SWB WS</u> No.1153, pp.C1/1-11 and No.1156, pp.C1-13. Also Luu (1982), pp.44-45 and White (1982a,b).

14. White (1982b), pp.2-3.

15. <u>Nhan Dan</u> 27/8/81, <u>SWB WS</u> No.1153, p.C8.

16. H 12/8/80, <u>SWB WS</u> No.1096, p.A31.

17. Le Duc Tho H 4-9/9/82, <u>SWB III</u> No.7144, p.C7.

18. For the resolution of the 6th Plenum of the CPV CC on Consumer Goods and Local Industries, see H 9, 10, and 11/10/79, <u>SWB III</u> No.6246, pp.C7-10 and No.6247, pp.C1-8; also the series of <u>Nhan Dan</u> editorials of this period.

19. H 7 and 8/7/80, <u>SWB III</u> No.6476, p.C3.

20. A detailed and relatively frank analysis of the origin of the new system, the problems of the old, and unresolved difficulties was provided by Le Duc Tho H 4-9/9/82, <u>SWB III</u> No.7144, pp.C1-20. This speech illustrates the compromise involved in the new policy not only within the

leadership, but for individuals. Le Duc Tho had argued in 1980 for increasing administrative control and economic management; see White (1982a), pp.23-24.

21. *Vietnam Courier*, *op. cit.*, pp. 12-13.

22. Ibid., p.13.

23. *Nhan Dan* 13/1/82, *SWB III* No.6933, p.C5.

24. 'There will be a residual number of labourers when the new contract system is applied.' Le Duc Tho, *op. cit.*, p.C15.

25. Ibid., p.C17.

26. Ibid., p.C10.

27. See White (1982a,b) and Le Duc Tho, *op. cit.*, p.C17, who expresses concern over the effect of work on the health of children.

28. Le Duc Tho, *op. cit.*, p.C12.

29. All of these points are from Le Duc Tho, pp.C6-20 and Nguyen Khac Vien in Gough (1982).

30. 'People went to work very late and would try to return home as early as possible, working four hours at most, only half the time required. Jobs which were difficult to check were done sloppily.' *Vietnam Courier*, *op. cit.*, p.12.

31. Le Duc Tho, *op. cit.*, p.C15.

32. Nguyen Khac Vien in Gough (1982), p.378. The neglect of handicrafts has been discussed earlier and is illustrated by Figure 4.

33. Le Duc Tho, *op. cit.*, p.C6. The assertion that this is also true in the USSR is made by Bradley and Clark (1972), pp.470-71.

Bibliography

The Bibliography is in two sections. The major part is a bibliography of agriculture in the Democratic Republic of Vietnam which contains virtually all of the material relevant to this subject in the English and French languages, as well as much of the material available in Vietnamese, outside of Vietnam. It thus constitutes a source of reference for other researchers.

The Supplementary Bibliography contains all the material referred to in this book which is not directly concerned with Vietnam. It includes theoretical works as well as material relating to the USSR and China.

1. Vietnam Bibliography

Adams, J., and Hancock, N. 1970. 'Land and Economy in Traditional Vietnam.' *Journal of Southeast Asian Studies*, September.

Ajalbert, J. 1912. *Les Nuages sur l'Indochine*. Paris: Louis-Michaud.

Anh Van and Roussel, J. 1947. *Mouvements Nationaux et Lutte des Classes au Vietnam*. Paris: Imprimerie Réaumur.

Annuaire Statistique de l'Indochine. Various years. See Service de la Statistique Générale de l'Indochine.

Bain, C.A. 1956. 'The History of Vietnam from the French Penetration to 1939.' Ph.D. thesis, American University.

Ball, W.M. 1952. *Nationalism and Communism in East Asia*. Melbourne: Melbourne University Press.

Bergman, A.E. 1975. *Women of Vietnam*. San Francisco: Peoples Press.

Bernard, F. 1901. *L'Indochine, Erreurs et Dangers: Un Programme*. Paris: Bibliothèque-

Charpentier.

Bernard, P. 1934. Le Problème Économique Indochinois. Paris: Nouvelles Editions Latines.

_____1937. Nouveaux Aspects du Problème Économique Indochinois. Paris: Fernand Sorbot.

Betts, R.F. 1961. Assimilation and Association in French Colonial Theory: 1890-1914. New York and London: Columbia University Press.

Bhaduri, A. 1979. Agricultural Cooperatives and Peasant Participation in the Socialist Republic of Vietnam. Geneva: International Labor Organization (ILO).

Boudarel, G. 1977. Giap. Paris: Editions Atlas.

British Broadcasting Corporation, Summary of World Broadcasts: Part V The Far East 1949-59 (SWB V); Part V The Far East Economic Supplement 1955-59 (SWB ES); Part III Far East 1959- (SWB III); Part III Far East Weekly Supplement 1959- (SWB WS).

Bui Huy Dap. 1967. 'Some characteristic features of rice growing in Vietnam.' Vietnamese Studies, No.13.

_____1971. 'Agronomic Research.' Vietnamese Studies, No.27.

Burchett, W.G. 1957. North of the 17th Parallel. Hanoi: Red River Publishing House.

_____1966. Vietnam North. New York: International Publishers.

Buttinger, J. 1958. The Smaller Dragon: A Political History of Vietnam. New York: Praeger.

_____1967. A Dragon Embattled: A History of Colonial and Post-Colonial Vietnam. New York: Praeger.

_____1968. Vietnam: A Political History. New York: Praeger.

Centre Tricontinental. 1979. La Politique Agraire au Nord Vietnam. Louvain: Centre Tricontinental.

Chaffard, G. 1964. Indochine: Dix Ans d'Indépendence. Paris: Calmann-Lévy.

Chaliand, G. 1968. The Peasants of North Vietnam. Harmondsworth: Allen Lane, The Penguin Press.

Charrière, J. 1965. 'Problèmes de la Construction du Socialisme en RDVN.' Economie et Politique, April.

____1966. 'Socialism in North Vietnam.' Monthly Review, February.

Chassigneux, E. 1912. L'Irrigation dans le Delta du Tonkin. Paris: Libraire Ch. Delgrave.

Chattopadhay, B. 1977. Water, Cereals and Economic Growth in South and East Asia in the Fifties and Sixties. New Delhi: Peoples Publishing House.

Chen, K.C. 1969. Vietnam and China, 1938-54. Princeton: Princeton University Press.

Cheshkov, M.A. 1969. 'La Classe Dirigeante du Vietnam Précolonial.' La Pensée, March-April.

Chesneaux, J. 1955. Contribution à l'Histoire de la Nation Vietnamienne. Paris: Editions Sociales.

____1966. The Vietnamese Nation: Contribution to a History. Sydney: Current Book Distributors.

____1968. Le Vietnam, Etudes de Politique et d' Histoire. Paris: Editions Maspéro.

Chesneaux, J., Boudarel, G., and Hémery, D. eds. 1971. Tradition et Révolution au Vietnam. Paris: Editions Anthropos.

Cole, A.B., ed. 1956. Conflict in Indochina and

International Repercussions: A Documentary History, 1945-55. Ithaca: Cornell University Press.

Cuong Pham. 1964 'Management Work in Agricultural Cooperatives,' Vietnamese Studies, No.2.

Delattre, J. 1961. 'L'Economie Vietnamienne au Début de son Premier Quinquennat.' Economie et Politique, June.

Devillers, P. 1952. Histoire du Vietnam de 1940 à 1952. Paris: Editions de Seuil.

Dao The Tuan. 1974. 'New Rice Strains.' Vietnamese Studies. No.38.

Doan Trong Truyen and Pham Thanh Vinh. 1964. Building an Independent National Economy in Vietnam. Hanoi: Foreign Languages Publishing House.

____1966. L'Edification d'une Economie Nationale Indépendente au Vietnam, 1945-65. Hanoi: Editions en Langues Etrangères.

Duiker, W.J. 1973. 'The Red Soviets of Nghe-Tinh: An Early Communist Rebellion in Vietnam.' Journal of Southeast Asian Studies, September.

____1975a. The Comintern and Vietnamese Communism. Southeast Asia Program, Monograph No.37. Athens, Ohio: Ohio University Centre for International Studies.

____1975b. The Rise of Nationalism in Vietnam, 1900-41. Ithaca: Cornell University Press.

____1975c. 'Building the United Front: The Rise of Communism in Vietnam: 1925-54.' In Communism in Indochina: New Perspectives, J.T. Zasloff and MacAllister Brown, eds. Lexington, Mass.: D.C. Heath.

____1981. The Communist Road to Power in Vietnam. Boulder: Westview Press.

Duong Binh. 1976. 'Water Control in Agriculture.' Vietnamese Studies, No.38.

Duong Hong Hien. 1974. 'Structuring of Crops.' Vietnamese Studies, No.38.

Dumarest, A. 1935. La Formation des Classes Sociales en Pays Annamite. Lyon: Imprimérie Ferréol.

Dumesnil, H. 1970. 'La Guerre et l'Economie de la République Démocratique du Vietnam.' Tiers Monde, April-September.

Dumont, R. 1935. La Culture du Riz dans le Delta du Tonkin. Paris: Société d'Editions Géographiques, Maritimes et Coloniales.

____1965a. 'Problèmes Agricoles au Nord Vietnam.' France-Asie, Autumn.

____1965b. Chine Surpeuplée, Tiers-Monde Affamé. Paris: Editions du Seuil.

____1973. Socialism and Development. London: Andre Deutsch.

Duncanson, D. 1968. Government and Revolution in Vietnam. New York: Oxford University Press.

Elliot, D.W.P. 1975. 'Political Integration in North Vietnam, The Cooperativisation Period.' In Communism in Indochina: New Perspectives, J.T. Zasloff and MacAllister Brown, eds. Lexington, Mass.: D.C. Heath.

____1976. 'Revolutionary Re-integration: A Comparison of the Foundation of Post-Liberation Political Systems in North Vietnam and China.' Ph.D. thesis, Cornell University.

____1980. 'Institutionalising the Revolution: Vietnam's Search for a Model of Development.' In Vietnamese Communism in Comparative Perspective, W.S. Turley. Boulder: Westview Press.

Ennis, T.E. 1956. French Policy and Developments in Indochina. Chicago: University of Chicago Press.

Fall, B.B. 1956. _The Viet-Minh Regime_. Ithaca: Cornell University Southeast Asia Programme and the Institute of Pacific Relations.

____1957. 'Crisis in North Vietnam.' _Far Eastern Survey_, January.

____1960. _Le Viet Minh: La Républic Démocratique du Vietnam, 1945-60_. Paris: Librairie Armand Colin.

____1961. _Street Without Joy: Insurgency in Indochina_. Harrisburg: Stackpole.

____1963a. _The Two Vietnams: A Political and Military Analysis_. New York: Praeger.

____1963b. 'A "Straight Zig-zag": The Road to Socialism in North Vietnam.' In _Communist Strategies in Asia, A Comparative Analysis of Governments and Parties_. A.D. Barnett, ed. New York: Praeger.

Feysal, P. de. 1933. _L'Endettement Agraire en Cochinchina_. Hanoi: Imprimerie d'Extrême-Orient.

Fitzgerald, F. 1973. _Fire in the Lake: The Vietnamese and the Americans in Vietnam_. New York: Vintage Books.

Food and Agricultural Organisation (FAO). 1948. _Yearbook of Agricultural Statistics_. Rome: FAO.

Fridland, V.M. 1961. _Priorada Severnogo V'etnama_ [The Physical Nature of North Vietnam]. Translation by Joint Publishers Research Services (JPRS). Moscow: Academy of Science USSR, Popular Science Series.

Ginsburgs, G. 1962 and 1963. 'Local Government and Administration in North Vietnam, 1945-54,' and 'Local Government and Administration in the Democratic Republic of Vietnam Since 1954,' Parts I and II. _China Quarterly_, Nos.10, 12, and 14.

Gittinger, J.P. 1959. 'Communist Land Policy in

North Vietnam.' Far Eastern Survey, August.

Gordon, A. 1971a. Economic Development and Socialism in North Vietnam. London: SOAS Left Group.

_____1971b. 'Debate on Agriculture in North Vietnam.' New Left Review, July/August.

_____1974. 'The "Green Revolution" in North Vietnam.' Journal of Contemporary Asia, Vol.4, No.1.

_____1977. 'Class Struggle on the Basis of Socialist Transition in North Vietnam.' Discussion Paper No.27, Economics Dept., University of Papua New Guinea.

_____1978a. 'Mistakes, Deviations or Class Struggle? North Vietnam's Land Reform, 1955-56.' Peasant Seminar Paper, Centre of International and Area Studies, University of London, January.

_____1978b. 'The Role of Class Struggle in North Vietnam.' Monthly Review, January.

_____1981. 'North Vietnam's Collectivisation Campaigns: Class Struggle, Production and the "Middle Peasant."' Journal of Contemporary Asia, Vol.2, No.1.

_____1982. 'The "End" of Class Struggle in Rural Vietnam Since 1975.' Paper presented to Transnational Institute (TNI) Seminar on Vietnam, June.

Goudal, J. 1938. Labour Conditions in Indochina. Geneva: International Labour Office.

Gough, K. 1978. 'The Green Revolution in South India and North Vietnam.' Bulletin of Concerned Asian Scholars. No.10.

_____1982. 'The Vietnamese Economy'--an interview with Nguyen Khac Vien. Journal of Contemporary Asia, Vol.12, No.3.

Gourou, P. 1931. Le Tonkin. Paris: Exposition

Coloniale Internationale.

____1936. Les Paysans du Delta Tonkinois: Etude de Géographie Humaine. Paris: Editions d'Art et d'Histoire.

____1945. Land Utilisation in French Indochina. Washington: Institute of Pacific Relations.

Gran, G. 1973. 'Vietnam and the Capitalist Route to Modernity: Village Cochinchina, 1880-1940.' Ph.D. thesis, Department of History, University of Wisconsin.

Hammer, E.J. 1955. The Struggle for Indochina. Stanford: Stanford University Press.

Hanoi--Publications without acknowledged authors. Unless otherwise stated all books are published by Foreign Languages Publishing House/ Editions en Langues Etrangères.

1955a. The Vietnamese Working Class in the Long and Hard War of Resistance.

1955b. Agrarian Reform Law.

1958. Economic Restoration and Cultural Development in the Democratic Republic of Vietnam, 1955-57.

1959. The Three Year Plan to Develop and Transform Economy and to Develop Culture.

1960a. Documents of the Third National Congress of the Vietnam Workers' Party.

1960b. The Democratic Republic of Vietnam.

1960c. Kinh Tế Việt Nam, 1945-60. [The Vietnamese Economy, 1945-60] (Nha Xuất Bản Sự Thật).

1960d. Breaking our Chains: Documents of the Vietnamese Revolution of August 1945.

1960e. 5 Năm Xây Dựng Kinh-Tế Và Văn Hóa, 1955-59. [5 Years of Building the Economy and Culture, 1955-59] Cục Thống

Kê Trung Ương.

1960f. <u>Thành Tích Sản Xuất Nông Nghiệp Trong 15 Năm Dưới Chế Độ Việt Nam Dân Chủ Cộng Hòa</u>. [The Results of Agricultural Production during the 15 Years of the DRV]. Bộ Nông Lâm.

1963a. 'Resolution of the 7th Plenum of the Vietnam Workers' Party,' <u>The Road to Happiness and Prosperity</u>.

1963b. <u>The Democratic Republic of Vietnam on the Road of Socialist Industrialisation</u>.

1963c. 'Resolution of the 5th Plenum of the Vietnam Workers' Party Central Committee on the Development of Agriculture,' <u>Offensive Against Poverty and Backwardness</u>.

1964a. <u>Communique of the 9th Plenum of the Central Committee of the Vietnam Workers' Party</u>.

1964b. 'Resolution of the 5th Plenum of the Vietnam Workers' Party Central Committee (July 1961) on the Development of Agriculture in the First Five-Year Plan (1961-1965),' <u>Vietnamese Studies</u>, No.2.

1964c. 'Ideological and Political Work in Agricultural Cooperatives.' <u>Vietnamese Studies</u>, No.2.

1964d. 'General Regulations Concerning Agricultural Producers' Lower-Type Cooperatives.' <u>Vietnamese Studies</u>, No.2.

1964e. 'Agricultural Statistics of the Democratic Republic of Vietnam.' <u>Vietnamese Studies</u>, No.2.

1964f. <u>Số Liệu Thống-Kê 1963</u>. [Statistical Material, 1963]. Nhà Xuất Bản Sự Thật.

1965. <u>A Heroic People: Memoirs From the Revolution</u>.

1967. 'Small Agricultural Mechanisation in the DRV.' <u>Vietnamese Studies</u>, No.13.

1969. *Role of the Vietnamese Working Class and Tasks of the Trade Unions at the Present Stage*.

1970a. *An Outline History of the Vietnam Workers' Party, 1930-70*.

1970b. 'Political Theses of the Indochinese Communist Party, 1930.' *Vietnamese Studies*, No.24.

1971. 'Constitution of High-Level Cooperative Farms.' *Vietnamese Studies*, No.27.

1974. *Vietnam: An Historical Sketch*.

1975. *The Democratic Republic of Vietnam*.

1976. *An Outline History of the Vietnam Workers' Party, 1930-75*.

1977a. *Fourth National Congress Documents*.

1977b. *Vietnam—Vers une Nouvelle Etape*.

1977c. *Kinh-Tế Và Văn Hóa Miền Bắc Xã Hội Chủ Nghĩa Việt-Nam, 1960-75*. [Economy and Culture in Socialist North Vietnam, 1960-75]. Tổng Cục Thống Kê.

1977d. 'Resolution of the 2nd Plenum (4th Congress) of the Central Committee of the Communist Party of Vietnam on Agricultural Development.' *Vietnamese Studies*, No.51.

1978. *30 Năm Phát Triển Kinh-Tế Và Văn Hóa*. [30 Years Economic and Cultural Development]. Nhà Xuất Bản Sự Thật.

Hémery, D. 1975. *Révolutionaires Vietnamiens et Pouvoir Colonial en Indochine*. Paris: François Maspéro.

Henry, Y. 1932. *L'Economie Agricole de l'Indochine*. Hanoi: Imprimerie d'Extrême Orient.

Hinton, W. 1968. *Fanshen: A Documentary of Revolution in a Chinese Village*. New York: Random House Vintage Books.

House Vintage Books.

Ho Chi Minh. 1962-64. Selected Works. 4 Vols. Hanoi: Foreign Languages Publishing House.

____1967. On Revolution. Selected Writings, 1920-66. New York: The New American Library.

____1969. Selected Articles and Speeches, 1920-67. London: Lawrence and Wishart.

____1977. Selected Writings (1920-1969). Hanoi: Foreign Languages Publishing House.

Hoai Thanh et al. 1962. Strong Wind. Hanoi: Foreign Languages Publishing House.

Hoang Van Chi, ed. 1958. The New Class in North Vietnam. Saigon: Cong Dan.

____1962. 'Collectivisation and Rice Production.' The China Quarterly, January-March.

____1964. From Colonialism to Communism: A Case History of North Vietnam. New York: Praeger.

Hodgkin, T. 1981. Vietnam: The Revolutionary Path. London: Macmillan.

Honey, P.J., ed. 1962. North Vietnam Today: Profile of a Communist Satellite. New York: Praeger.

____1966. Communism in North Vietnam. Cambridge, Mass.: MIT Press.

Huynh Kim Khanh. 1982. Vietnamese Communism, 1925-45. Ithaca: Cornell University Press.

Huynh Ngoc Buu. 1971. 'Pisciculture.' Vietnamese Studies, No.27.

Isoart, P. 1961. Le Phénomène National Vietnamien: De l'Indépendance Unitaire à l'Indépendance Fractionnée. Paris: Librairie Générale de Droit et de Jurisprudence.

Jacoby, E.H. 1961. Agrarian Unrest in Southeast Asia. Bombay: Asia Publishing House.

Jones, P.H.M. 1960. 'Industry of North Vietnam.' Far Eastern Review.

Karamyshev, V.P. 1959. Sel'skoe Khozyaystvo Demokraticheskoy Respubliki V'etnam. [Agriculture in the Democratic Republic of Vietnam]. Moscow: State Publishing House of Agricultural Literature. English translation, JPRS, 4363, 1961.

Kaye, W. 1962. 'A Bowl of Rice Divided: The Economy of North Vietnam.' China Quarterly, January-March.

Kresser, P. 1935. La Commune Annamite en Cochinchine. Paris: F. Loviton.

Lacouture, J. 1966. Vietnam: Between the Two Truces. New York: Random Books.

____1968. Ho Chi Minh: A Political Biography. New York: Random Books.

Lancaster, D. 1961. The Emancipation of French Indochina. London and New York: Oxford University Press.

Lavallée, L. 1971. L'Economie du Nord Vietnam, 1960-70. Paris: Centre d'Etudes et de Recherches Marxistes.

Le Chau. 1966. Le Vietnam Socialiste. Paris: François Maspéro.

Le Duan. 1965. On the Socialist Revolution in Vietnam. 3 Vols. Hanoi: Foreign Languages Publishing House.

____1971. The Vietnamese Revolution: Fundamental Problems, Essential Tasks. Hanoi: Foreign Languages Publishing House.

____1974. Quelques Tâches Actuelles. Hanoi: Editions en Langues Etrangères.

____1977. Selected Writings. Hanoi: Foreign Languages Publishing House.

Le Duan and Pham Van Dong. 1975. Towards Large-Scale Socialist Agricultural Production. Hanoi: Foreign Languages Publishing House.

Le Thanh Khoi. 1955. Le Vietnam: Histoire et Civilisation. Paris: Editions de Minuit.

____, ed. 1978. Socialisme et Développement au Vietnam. Paris: Presses Universitaires de France.

Le Vinh. 1962. 'Problèmes de l'Industrialisation.' La Nouvelle Critique, March.

Limbourg, M. 1956. L'Economie Actuelle du Vietnam Democratique. Hanoi: Editions en Langues Etrangères.

MacAllister, J.T., Jr. 1969. Vietnam: The Origins of Revolution. New York: Alfred A. Knopf.

MacAllister, J.T., Jr. and Mus, P. 1970. The Vietnamese and Their Revolution. New York: Harper and Row.

Mai Thi Tu and Le Thi Nham Tuyet. 1978. Women in Vietnam. Hanoi: Foreign Languages Publishing House.

Mao Tse Tung, 1971. Selected Readings from the Works of Mao Tse Tung. Peking: Foreign Languages Press.

Marr, D. 1971. Vietnamese Anti-Colonialism, 1885-1925. Berkeley and Los Angeles: University of California Press.

____ 1981. Vietnamese Tradition on Trial, 1920-45. Berkeley and Los Angeles: University of California Press.

Martin, F. 1948. Heures Tragiques au Tonkin: 9 mars 1945 - 18 mars 1946. Paris: Berger-Levraut.

Masson, A. 1950. Histoire de l'Indochine. Paris: Presses Universitaire de France.

____1960. Histoire du Vietnam. Paris: Presses Universitaires de France.

Mau, M.P. 1977. 'The Political Evolution of the Village Commune in North Vietnam, 1802-1970.' Ph.D. thesis, University of Pennsylvania.

Moise, E.E. 1976. 'Land Reform and Land Reform Errors in North Vietnam.' Pacific Affairs, Spring.

____1977. 'Land Reform in China and North Vietnam: Revolution at the Village Level.' Ph.D. thesis, University of Michigan.

____1980. '"Class-ism" in North Vietnam, 1953-56.' In Vietnamese Communism in Comparative Perspective, W.S. Turley, ed. Boulder: Westview Press.

Munholland, J.K. 1975. 'The French Response to the Vietnamese Nationalist Movement 1905-14.' Journal of Modern History.

Murray, M.J. 1980. The Development of Capitalism in Colonial Indochina, 1870-1940. Berkeley: University of California Press.

Murti, B.S.N. 1964. Vietnam Divided: The Unfinished Struggle. New York: Asia Publishing House.

Mus, P. 1949. 'The Role of the Village in Vietnamese Politics.' Pacific Affairs, September.

____1950. Vietnam: Sociologie d'une Guerre. Paris: Editions du Seuil.

Ng Shui Meng. 1974. The Population of Indochina. Singapore: Singapore Institute of South East Asian Studies.

Ngo Tat To. 1959. Quand la Lampe s'Eteint. Hanoi: Editions en Langues Etrangères.

Ngo Vinh Long. 1973. Before the Revolution: The Vietnamese Peasants Under the French. Cambridge, Mass.: MIT Press.

____1978. 'The Indochinese Communist Party and Peasant Rebellion in Central Vietnam, 1930-1.' Bulletin of Concerned Asian Scholars, Vol.10, No.4.

Nguyen Chi Thanh. 1963a. 'Situation de l'Agriculture Nord Vietnamienne.' Tiers Monde, January-June.

____1963b. 'On Crop Management' in Offensive Against Poverty and Backwardness. Hanoi: Foreign Languages Publishing House.

Nguyen Duy Trinh. 1958. 'Government Report on the Three-Year Plan of North Vietnam, 1958-60.' Nhân Dân, 11-13 December. JPRS translation 632-D Washington, D.C., 1959.

Nguyen Huu Khang. 1946. La Commune Annamite, Etude Historique, Juridique et Economique. Paris: Tepac.

Nguyen Khac Vien. 1964. 'Water, Rice and Men.' Vietnamese Studies, No.2.

____, ed. 1965. Vietnam Today. Hanoi: Foreign Languages Publishing House.

____1970. Experiences Vietnamiennes. Paris: Editions Sociales.

____1974. Histoire du Vietnam. Paris: Editions Sociales.

____1975a. The Long Resistance 1858-1975. Hanoi: Foreign Languages Publishing House.

____1975b. Tradition and Revolution in Vietnam. Berkeley, California: Indochina Resource Center.

____, ed. 1976. The Socialist Republic of Vietnam Structure and Basis. Hanoi: Foreign Languages Publishing House.

____, ed. 1978. The Socialist Republic of Vietnam: Basic Data. Hanoi: Foreign Languages Publishing House.

Nguyen Khac Vien et al. 1967. 'The Agricultural Cooperation Policy.' _Vietnamese Studies_, No.13.

Nguyen Kien Giang. 1960. _Les Grandes Dates du Parti de la Classe Ouvrière du Vietnam_. Hanoi: Editions en Langues Etrangères.

Nguyen Lai Vien. 1967. 'The Battle of Rice.' _Vietnamese Studies_, No.13.

Nguyen Lao Dong. 1971. 'Animal Husbandry.' _Vietnamese Studies_. No.27.

Nguyen Ngoc Luu. 1982. _Institutional Factors and Technological Innovation in North Vietnamese Agriculture_. Working Paper, Technology and Employment Programme. Geneva: ILO.

Nguyen Ngoc Minh. 1965. 'The Birth of an Independent Currency.' _Vietnamese Studies_, No.7.

Nguyen Thi. 1976. _A Village Called Faithfulness._ Hanoi: Foreign Languages Publishing House.

Nguyen Tien Hung. 1977. _Economic Development of Socialist Vietnam, 1955-80_. New York: Praeger.

Nguyen Van Vinh. 1961. _Les Réformes Agraires au Vietnam_. Brussels: Libraires Universitaires Urstpruyst.

Nguyen Xuan Lai. 1957. 'La Reconstruction Economique de la République Démocratique du Vietnam.' _Economie et Politique_, August-September.

____1967. 'The Family Economy of Cooperative Farmers.' _Vietnamese Studies_, No.13.

____1968. 'The Economy of North Vietnam Facing the Trial of War.' _Vietnamese Studies_, No.17.

____1971. 'Interdependence Between Agriculture and Industry.' _Vietnamese Studies_, No.27.

____1976. 'The First Resistance (1945-54).' _Vietnamese Studies_, No.44.

Vietnamese Studies, No.44.

____, ed. 1977. 'The Management of Cooperatives.' Vietnamese Studies, No.51.

Nguyen Xuan Lai and Vu Quoc Tuan. 1976. 'The Second Resistance (1965-72).' Vietnamese Studies, No.44.

Nhân Dân. [The People's Daily]. Hanoi.

Osborne, M.E. 1969. The French Presence in Cochinchina and Cambodia: Rule and Response 1859-1905. Ithaca: Cornell University Press.

____1974. 'Continuity and Motivation in the Vietnamese Revolution: New Light from the 1930s.' Pacific Affairs, Spring.

Paige, J. 1975. Agrarian Revolution: Social Movements and Export Agriculture in the Underdeveloped World. New York: Free Press.

Pham Cuong. 1971. 'The Training of Technicians for Agricultural Cooperatives.' Vietnamese Studies, No.27.

____1975. 'Thirty Years of Agricultural Development.' Vietnam Courier, No.42.

Pham Cuong and Nguyen Van Ba. 1976. Revolution in the Village Nam Hong, 1945-75. Hanoi: Foreign Languages Publishing House.

Pham Van Dong. 1970. 25 Years of National Struggle and Construction. Hanoi: Foreign Languages Publishing House.

Pike, D. 1972. 'Report on North Vietnam.' Asian Survey, January.

____1978. History of Vietnamese Communism, 1925-76. Stanford, California: Hoover Institution Press.

Popkin, S.L. 1979. The Rational Peasant: The Political Economy of Rural Society in Vietnam. Berkeley and Los Angeles: University of California Press.

Porter, D.G. 1972. North Vietnam's Land Reform. Interim Report No.2, International Relations of East Asia Project. Ithaca: Cornell University.

____1973. 'The Myth of the Bloodbath: North Vietnam's Land Reform Reconsidered.' Bulletin of Concerned Asian Scholars, September.

____1976. 'Proletariat and Peasantry in Early Vietnamese Communism.' Asian Thought and Society, December.

____, ed. 1979. The Definitive Documentation of Human Decisions. London: Heyden.

Post, K.W.P. Revolution, Nationalism and Socialism in Vietnam. The Hague, forthcoming.

Rambo, A.T. 1972. 'A Comparison of Peasant Social Systems of Northern and Southern Vietnam: A Study of Ecological Adaptation, Social Succession and Cultural Evolution.' Ph.D. thesis, Department of Anthropology, University of Hawaii.

Robequain, C. 1929. Le Thanh Hoa: Etude Géographique d'une Province Annamite. 2 Vols. Paris: Publications de l'Ecole Française d'Extrême-Orient.

____1944. The Economic Development of French Indochina. London: Oxford University Press.

Roberts, S.H. 1929. History of French Colonial Policy: 1870-1925. 2 Vols. London: P.S. King.

Robertson, C.J. 1936. 'The Rice Export From Burma, Siam and French Indochina.' Pacific Affairs, June.

Roubaud, L. 1931. Vietnam: La Tragédie Indochinoise. Paris: Libraire Valois.

Rouilly, M. 1929. La Commune Annamite. Paris: Les Presses Modernes.

Rousset, P. 1975. Le Parti Communiste Vietnamien.

Rousset, P. 1975. *Le Parti Communiste Vietnamien*. Paris: Maspéro.

____1978. *Communisme et Nationalisme Vietnamien*. Paris: Editions Galilée.

SWB (Summary of World Broadcasts) see British Broadcasting Corporation.

Sacks, I.M. 1959. 'Marxism in Vietnam.' In *Marxism in Southeast Asia: A Study of Four Countries*. F.N. Trager, ed. Stanford: Stanford University Press.

Sainteny, J. 1953. *Histoire d'une Paix Manquée*. Paris: Amiot Dumont.

Sampson, C. 1975. 'Nationalism and Communism in Vietnam, 1925-31.' Ph.D. thesis, University of California, Los Angeles.

Sansom, R. 1970. *The Economics of Insurgency in the Mekong Delta of Vietnam*. Cambridge, Mass.: MIT Press.

Scott, J.C. 1976. *The Moral Economy of the Peasant: Rebellion and Subsistence in Southeast Asia*. New Haven and London: Yale University Press.

Service de la Statistique Générale de l'Indochine. *Annuaire Statistique de l'Indochine*. Various years. Direction des Services Economiques.

Shabad, T. 1958. 'Economic Developments in North Vietnam.' *Pacific Affairs*, March.

Smith, H.H. et al. 1967. *Area Handbook for North Vietnam*. Washington, D.C.: US Gov't. Printing Office.

Thanh Luong. 1965. *A Short History of Vietnam*. Hanoi: Foreign Languages Publishing House.

Thompson, V. 1937. *French Indochina*. London: Allen and Unwin.

____1947. *Labour Problems in Southeast Asia*.

London: Oxford University Press.

Tran Dong Khoa. 1964. 'Water Conservancy in North Vietnam.' Vietnamese Studies, No.2.

Tran Huy Lieu. 1960. Les Soviets du Nghe-Tinh (de 1930-1) au Vietnam. Hanoi: Editions en Langues Etrangères.

Tran Ngoc Bich. 1972. 'Stratégie de Développement et Evolution du Cadre Socio-Economique au Nord Vietnam.' Civilisations, No.1.

Tran Nhu Trang. 1972. 'The Transformation of the Peasantry in North Vietnam.' Ph.D. thesis, University of Pittsburgh.

Tran Phuong. 1964. 'The Agrarian Peasant Question in the Vietnamese Revolution.' Vietnamese Studies, No.4.

____1965. 'The Land Reform.' Vietnamese Studies. No.7.

____, ed. 1968. Cách Mạng Ruộng Đất ở Việt-Nam. [Agrarian Revolution in Vietnam]. Hanoi: Nhà Xuất Bản Khoa Học Xã Hoi.

Tran Quan Thuyet and Dao The Tuan. 1974. 'Azolla: A Green Compost.' Vietnamese Studies, No.38.

Trinh Van Tinh. 1964. 'Renovation of Agricultural Techniques in North Vietnam.' Vietnamese Studies, No.2.

Truong Buu Lam. 1967. Patterns of Vietnamese Response to Foreign Intervention: 1858-1900. Mono. No.11. New Haven, Conn.: Yale Southeast Asia Studies.

Truong Chinh and Vo Nguyen Giap. 1974. The Peasant Question (1937-8). Translated by C.P. White. Southeast Asia Programme, Data Paper No.94. Ithaca: Cornell University Press.

Truong Chinh. 1955. Pour la Réalisation de la Réforme Agraire. Hanoi: Editions en Langues Etrangères.

____1959. _Resolutely Taking the North Vietnam Countryside to Socialism Through Agricultural Cooperation_. Hanoi: Foreign Languages Publishing House.

____1963a. _Primer for Revolt: The Communist Takeover in Vietnam_. New York: Praeger.

____1963b. _March Ahead Under the Party's Banner_. Hanoi: Foreign Languages Publishing House.

____1969. _Forward Along the Path Charted by Karl Marx_. Hanoi: Foreign Languages Publishing House.

____1970. 'Weakness, Shortcomings and Mistakes in Agricultural Cooperatives.' In _Vietnam: Documents and Research Notes_. Saigon: United States Mission in Vietnam.

____1977. _Selected Writings_. Hanoi: Foreign Languages Publishing House.

Truong Van Ba. 1962. 'De la Réforme Agraire à la Coopération.' _La Nouvelle Critique_, January-March.

Turley, W.S. 1972a. 'Army, Party and Society in the Democratic Republic of Vietnam.' Ph.D. thesis, University of Washington.

____1972b. 'The Democratic Republic of Vietnam Since the Death of Ho Chi Minh: The Politics of a Revolution in Transition.' In _Indochina in Conflict_, J.T. Zasloff and A.E. Goodman. Lexington, Mass.: D.C. Heath.

____1975. 'Urbanisation in War: Hanoi, 1946-73.' _Pacific Affairs_, Fall.

____, ed. 1980. _Vietnamese Communism in Comparative Perspective_. Boulder: Westview Press.

Turner, R.F. 1975. _Vietnamese Communism: Its Origins and Development_. Stanford, California: Hoover Institution Press.

United States Congress, Senate. 1973. Committee on the Judiciary. _The Human Cost of Commun-_

ism in Vietnam - II: The Myth of No Bloodbath. Hearing before Subcommittee to investigate the Administration of the Internal Security Act and other Internal Security Laws, 93rd Congress, 1st Session, January 5, 1973. Washington, D.C.: US Gov't. Printing Office.

United States Department of Agriculture. 1950. The Agriculture of French Indochina. Prepared for the Economic Cooperation Administration by the Office of Foreign Agricultural Relations of the US Dept. of Agriculture. Washington, D.C.: US Gov't. Printing Office.

United States Mission in Vietnam. 1969. The New Statute on Agricultural Cooperation - in the Context of the Agricultural Policy Debate in North Vietnam. Saigon: US Mission in Vietnam.

Van Dyke, J.M. 1972. North Vietnam's Strategy for Survival. Palo Alto: Pacific Books.

Viet Chung. 1974. 'Industrial Crops.' Vietnamese Studies, No.38.

Viollis, A. 1935. Indochine S.O.S. Paris: Gallimard.

Vo Huy Bang. 1967. 'The Purchase of Farm Products by the State and the Price Problem.' Vietnamese Studies, No.13.

Vo Nhan Tri. 1960. 'La Politique Agraire du Nord Vietnam.' Tiers Monde, July-September.

____ 1967. Croissance Economique de la République Démocratique du Viet-Nam (1945-65). Hanoi: Editions en Langues Etrangères.

Vo Nguyen Giap. 1962. People's War, People's Army. New York: Praeger.

Vu Do Thin. 1954. Evolution Economique du Vietnam. Paris: Librairie Générale de Droit et de Jurisprudence.

White, C.P. 1970. 'Land Reform in North Vietnam.' Agency for International Development Spring Review, June.

____1974. 'The Vietnamese Revolutionary Alliance: Intellectuals, Workers and Peasants.' In *Peasant Rebellion and Communist Revolution in Asia*, J.W. Lewis, ed. Stanford: Stanford University Press.

____1978. 'Revolutionary Transformation in Village Leadership in the Democratic Republic of Vietnam.' Peasant Seminar Paper, Centre of International and Areas Studies, University of London.

____1979. 'The Peasants and the Party in the Vietnamese Revolution.' In *Peasants and Politics: Grass Roots Reaction to Change in Asia*, D.B. Miller, ed. London: Edward Arnold.

____1982a. *Debates in Vietnamese Development Policy*. Discussion Paper 171, Brighton, England: Institute of Development Studies, University of Sussex.

____1982b. 'Reforming Relations of Production, Family and Cooperative in Vietnamese Agricultural Policy.' Paper presented to Transnational Institute Seminar on Vietnam, June.

Woodside, A. 1970. 'Decolonisation and Agricultural Reform in Northern Vietnam.' *Asian Survey*, August.

____1971. *Vietnam and the Chinese Model: A Comparative Study of the Vietnamese and Chinese Government in the First Half of the Nineteenth Century*. Cambridge, Mass.: Harvard University Press.

____1976. *Community and Revolution in Modern Vietnam*. Boston: Houghton Mifflin.

2. Supplementary Bibliography

Baran, P.A. 1973. *The Political Economy of Growth*. Harmondsworth: Penquin.

Bradley, M.F. and Clark, M.G. 1972. 'Supervision

and Efficiency in Socialised Agriculture.' Soviet Studies, Vol.23, No.3.

Bukharin, N.I. 1982. Selected Writings on the State and the Transition to Socialism. Edited by Richard B. Day. New York: M.E. Sharpe.

Chayanov, A.V. 1966. The Theory of Peasant Economy. Illinois: Irwin.

Dobb, M. 1966. Soviet Economic Development Since 1917. London: Routledge and Kegan Paul.

Eckstein, A. 1961. 'The Strategy of Economic Development in Communist China.' American Economic Review, May.

Ellman, M. 1975. 'Did the Agricultural Surplus Provide the Resources for the Increase in Investment in the USSR During the First Five Year Plan?' Economic Journal, Vol.85.

____1978. 'On a Mistake of Preobrazhensky and Stalin.' Journal of Development Studies, Vol.14, No.4.

Erlich, A. 1950. 'Preobrazhensky and the Economics of Soviet Industrialisation.' Quarterly Journal of Economics, February.

____1967. The Soviet Industrialisation Debate. Cambridge, Mass.: Harvard University Press.

Gray, J. 1970. 'The High Tide of Socialism in the Chinese Countryside.' In Studies in the Social History of China and South-East Asia, J. Ch'en and N. Tarling, eds. London: Cambridge University Press.

Harrison, M. 1975. 'Chayanov and the Economics of the Russian Peasantry.' Journal of Peasant Studies, Vol.2, No.4.

Hill, I. 1975. 'The "Private Plot" in Soviet Agriculture.' Journal of Peasant Studies, July.

Hobsbawm, E.J., et al., eds. 1980. Peasants in History, Essays in Honour of Daniel Thorner.

Delhi: Oxford University Press.

Hsieh, S.C. 1971. 'Taiwan's Model of Agricultural Progress.' In _Agrarian Policies and Problems in Communist and Non-Communist Countries_, W.A. Douglas Jackson, ed. Seattle: University of Washington Press.

Ishikawa, S. 1967. _Economic Development in Asian Perspective_. Tokyo: Institute of Economic Research, Hototsubashu University.

Kahan, A. 1964. 'The Collective Farm System in Russia: Some Aspects of its Contribution to Soviet Economic Development.' In _Agriculture in Economic Development_, C. Eicher and L. Witt, eds. New York: McGraw-Hill.

Kautsky, K. 1900. _La Question Agraire_. Paris: Bibliothèque Socialiste Internationale 2.

Kerblay, B. 1966. 'A.V. Chayanov: Life, Career, Works.' In _The Theory of the Peasant Economy_, A.V. Chayanov. Illinois: Irwin.

____1971. 'Chayanov and the Theory of the Peasantry as a Specific Type of Economy.' In _Peasants and Peasant Societies_, T. Shanin, ed. Harmondsworth: Penguin.

____1980. 'Peasant Family Economy in the USSR Today.' In _Peasants in History_, E.J. Hobsbawm et al., eds. Delhi: Oxford University Press.

Kitching, G. 1982. _Development and Underdevelopment in Historical Perspective_. London: Methuen.

Lewin, M. 1980. 'The Kolkhoz and the Russian Muzhik.' In _Peasants in History_, E.J. Hobsbawm et al., eds. Delhi: Oxford University Press.

Limqueco, P. and McFarlane, B. 1980. 'Problems of Economic Planning for Underdeveloped Socialist Countries.' In _Planning Industrial Development_, D.F. Walker. London: John Wiley.

Littlejohn, G. 1973a. 'The Peasantry and the Russian Revolution.' _Economy and Society_, Vol.2.

____1973b. 'The Russian Peasantry: A Reply to Teodor Shanin.' _Economy and Society_, Vol.2.

Mandel, E. 1968. _Marxist Economic Theory_. London: Merlin.

Millar, J.R. 1970a. 'A Reformulation of A.V. Chayanov's Theory of the Peasant Economy.' _Economic Development and Cultural Change_, Vol.18, No.2.

____1970b. 'Soviet Rapid Development and the Agricultural Surplus Hypothesis.' _Soviet Studies_, Vol.22, No.1.

____1971. 'The Agricultural Surplus Hypothesis: A Reply to Alec Nove.' _Soviet Studies_, Vol.23, No.4.

Nolan, P. 1976. 'Collectivisation in China: Some Commparisons with the USSR.' _Journal of Peasant Studies_, Vol.3, No.2.

Nove, A. 1965. 'Introduction.' In _The New Economics_, E. Preobrazhensky. Oxford: Clarendon Press.

____1971. 'The Agricultural Surplus Hypothesis: A Comment on James R. Millar's Article.' _Soviet Studies_, Vol.22, No.3.

Paine, S. 1976. 'Balanced Development: Maoist Conception and Chinese Practice.' _World Development_, Vol.4, No.4.

Patnaik, U. 1976. 'Class Differentiation Within the Peasantry.' _Economic and Political Weekly_, September.

Perkins, D.H. 1969. _Agricultural Development in China 1368-1968_. Edinburgh: Edinburgh University Press.

Preobrazhensky, E. 1965. _The New Economics_. Oxford: Clarendon Press.

Riskin, C. 1971. 'Small Industry and the Chinese Model of Development.' China Quarterly, No.46.

Shanin, T. 1972. The Awkward Class: Political Sociology of the Peasantry in a Developing Society: Russia 1910-25. Oxford: Oxford University Press.

____1973. 'Gary Littlejohn's Review of T. Shanin, the Awkward Class.' Economy and Society, Vol.2.

____1980. 'Measuring Peasant Capitalism.' In Peasants in History, E.J. Hobsbawm et al, eds. Delhi: Oxford University Press.

Tang, A.M. 1968 'Policy and Performance in Agriculture.' In Economic Trends in Communist China, A. Eckstein, W. Galenson and T.C. Liu, eds. Edinburgh: Edinburgh University Press.

Thorner, D. 1966. 'Chayanov's Concept of Peasant Economy.' In The Theory of Peasant Economy. A.V. Chayanov. Illinois: Irwin.

Wadekin, K-E. 1971. "Kolkhoz, Sovkhoz and Private Production in Soviet Agriculture.' In Agrarian Policies and Problems in Communist and Non-Communist Countries, W.A. Douglas Jackson, ed. Seattle: University of Washington Press.

____1973. The Private Sector in Soviet Agriculture. Berkeley: University of California Press.

Walker, K.R. 1965. Planning in Chinese Agriculture. London: Frank Cass.

____1966. 'Collectivisation in Retrospect: "The Socialist High Tide" of Autumn 1955 - Spring 1956.' China Quarterly, No.26.

____1968. 'Organisation of Agricultural Production.' In Economic Trends in Communist China, A. Eckstein, W. Galenson and T.C. Liu, eds. Edinburgh: Edinburgh University Press.

Edinburgh: Edinburgh University Press.

Witfogel, K.A. 1971. 'Communist and Non-Communist Agrarian Systems, With Special Reference to the USSR and Communist China: A Comparative Approach.' In _Agrarian Policies and Problems in Communist and Non-Communist Countries_, W.A. Douglas Jackson, ed. Seattle: University of Washington Press.

THE AUTHOR

ANDREW VICKERMAN was educated at Cambridge University in England and has taught at the University of East Anglia and the University of Papua New Guinea.

He has left (temporarily?) the hallowed halls of academia and is currently Economic Advisor to the Prime Minister of Papua New Guinea.

YALE UNIVERSITY SOUTHEAST ASIA STUDIES
Box 13A Yale Station
New Haven, Connecticut 06520

RECENT PUBLICATIONS

Film on Indonesia, by Toby Alice Volkman. Catalogue, ix, 52pp. $4.00. A vlauable resource for teachers and students of Indonesian studies, anthropology, ethnographic film, and many other subjects embraced by the 87 films reviewed. (1985)

Manila 1900-1941: Social Change in a Late Colonial Metropolis by Daniel F. Doeppers. Mono.#27. ix, 194pp. $14.00. With scholarly text, copious notes, and solid statistics, this work explores the subject of the changing face of society in colonial Manila during the early 20th century. (1984)

Centers, Symbols, and Hierarchies: Essays on the Classical States of Southeast Asia, edited by Lorraine Gesick with Foreword by Clifford Geertz, Mono.#26. x, 241pp. $14.00. A collection of six essays which raises important theoretical issues as it examines the early Southeast Asian states of Burma, Cambodia, Java, Sulawesi, and Thailand. (1983)

Revolution and its Aftermath in Kampuchea: Eight Essays, edited by David P. Chandler and Ben Kiernan. Mono.#25. x, 319pp. $14.00. Provides the reader with a first-hand picture of the grim recent history of Kampuchea - of politics and life during the period 1975-1979 when the Khmer Rouge under their leader Pol Pot were in power. (1983)

Moral Order and the Question of Change: Essays on Southeast Asian Thought, edited by David K. Wyatt and Alexander Woodside. Mono.#24. vii, 413pp. $16.00. Scholars interested in the history of ideas, social history, or Southeast Asian literature will find much of value in this volume which contains seven essays covering Siam, Cambodia, Burma, Vietnam, Java, and the Philippines. (1982)

Peasant Politics and Religious Sectarianism: Peasant and Priest in the Cao Dai in Viet Nam, by Jayne S. Werner. Mono. #23. iv, 123pp. $10.50. A thought-provoking study of the history of the Cao Dai sect in Viet Nam from its founding in Saigon in 1925. (1981)

Southeast Asia Under Japanese Occupation, edited by Alfred W. McCoy. Mono.#22. vi, 250pp. $14.00. Nine essays which reassess the thesis of the late Harry Benda that the "Japanese interregnum" formed a distinct and decisive epoch in Southeast Asian history. (1980)

ADDITIONAL PUBLICATIONS CURRENTLY AVAILABLE

Monograph Series

21 Larkin, John A., ed., Perspectives on Philippine Historiography: A Symposium (1979), iv, 74 pages, with tables and figures, $9.50.

20 Cushner, Nicholas P., S.J., Landed Estates in the Colonial Philippines (1976), x, 146 pages, with appendices, tables, chart, maps, glossary, bibliography, index, $11.50.

19 Liddle, R. William, ed., et al., Political Participation in Modern Indonesia (1973), x, 206 pages, with tables, $9.50.

17 Reinhardt, Jon M., Foreign Policy and National Integration: The Case of Indonesia (1971), vi, 230 pages, with maps, chart, tables, $8.50.

16 Simkins, Paul D. and Frederick L. Wernstedt, Philippine Migration: The Settlement of the Digos-Padada Valley, Davao Province (1970), x, 147 pages, with maps, graphs, tables, photographs, $8.50.

15 Bennett, Paul J., Conference under the Tamarind Tree: Three Essays in Burmese History (1971), viii, 153 pages, with maps, bibliography, appendices, $8.25.

14 Bellows, Thomas J., The People's Action Party of Singapore: Emergence of a Dominant Party System (1970), xii, 195 pages, with appendices, $8.25.

2 McHale, Thomas R. and Mary C. McHale, Early American-Philippine Trade: The Journal of Nathaniel Bowditch in Manila, 1796 (1962), viii, 63 pages, $4.75.

Translation Series

7 Hendon, Rufus S., Six Indonesian Short Stories (1980), viii, 123 pages, $5.50.

Bibliography Series

11 Conklin, Harold C., Ifugao Bibliography (1968), vi, 75 pages, $5.50.

10 Hicks, George L. and Geoffrey McNicoll, The Indonesian Economy, 1950-1967: Bibliographic Supplement (1968), xii, 211 pages, with author index, subject index, list of serials, $6.75.

9 Hicks, George L. and Geoffrey McNicoll, The Indonesian Economy, 1950-1965: A Bibliography (1967), x, 248 pages, with author index, subject index, list of serials, $5.75.

FORTHCOMING PUBLICATIONS

Monograph Series

Philippine Colonial Democracy, edited by Ruby Paredes with chapters by Michael Cullinane, Alfred W. McCoy, Glenn May, and Ruby Paredes.

Pol Pot Plans the Future: Confidential Leadership Documents from Democratic Kampuchea, 1976-77, translated and edited by Chanthou Boua, David P. Chandler, and Ben Kiernan.

From Class to Culture: Social Conscience in Malay Novels since Independence by David J. Banks.

Indonesian Economic History of the Dutch Colonial Era, edited by Anne Booth, William J. O'Malley, and Anna Weidemann.

The Development of Capital and the Role of the State in Thailand, by Kevin Hewison.

Malaysia's National Language Policy and the Court System, by Richard Mead.

INQUIRIES AND ORDERS SHOULD BE ADDRESSED TO:

Yale University
Southeast Asia Studies
Box 13A Yale Station
New Haven, Connecticut 06520

For non-institutional orders, please enclose remittance adding $1.00 for postage and handling for the initial title and $.50 for each additional title. Please made check or money order payable to: Yale Univ. Southeast Asia Studies.